Robert von Lendenfeld

Descriptive Catalogue of the Medusæ of the Australian Seas

In two parts: Part I. Scyphomedusæ. Part II. Hydromedusæ

Robert von Lendenfeld

Descriptive Catalogue of the Medusæ of the Australian Seas
In two parts: Part I. Scyphomedusæ. Part II. Hydromedusæ

ISBN/EAN: 9783337384159

Printed in Europe, USA, Canada, Australia, Japan

Cover: Foto ©berggeist007 / pixelio.de

More available books at **www.hansebooks.com**

THE AUSTRALIAN MUSEUM.

DESCRIPTIVE CATALOGUE

OF THE

MEDUSÆ OF THE AUSTRALIAN SEAS.

IN TWO PARTS:

PART I.—SCYPHOMEDUSÆ.
PART II.—HYDROMEDUSÆ.

BY

R. VON LENDENFELD, PH.D.

PRINTED BY ORDER OF THE TRUSTEES.

E. P. RAMSAY, LL.D., F.R.S.E., Curator.

SYDNEY: CHARLES POTTER, GOVERNMENT PRINTER.

1887.

PART I.

SCYPHOMEDUSÆ.

CORRECTION.

The Plate referred to on page 4 of Part I, unfortunately cannot be re-produced; the work is, therefor, issued without it. The original painting is in the possession of the Linnean Society of N.S.W.

For *Cuidaria*, wherever occurring, read *Cnidaria*.

THE SCYPHOMEDUSÆ OF THE AUSTRALIAN SEAS;

By R. v. LENDENFELD, Ph.D.

INTRODUCTION.

THE difficulty connected with the preservation of these beautiful animals has been a great obstacle in the way of a thorough knowledge of them.

A number of species have been described from Australia in narratives of voyages. Some of these were figured from life on the ship. These figures are the only true guide to the identification of old descriptions with the forms we meet with. The descriptions given by the older authors are often short diagnoses, and therefore utterly useless; even some of the more recently described forms appear doubtful.

Haeckel (Das System der Medusen, 1879) has described a number of Australian species from spirit specimens, which were mostly very defective, so that many of his forms appear doubtful. It is by no means unlikely that some of his species are only the young stages of others. (Claus, Untersuchungen über die Organisation und Entwickelung der Medusen, 1883.) On the whole our knowledge of the Australian Scyphomedusæ was not at all in proportion to the great abundance, variety, and beauty of the Australian jelly-fish, when I, four years ago, commenced my studies of the Australian Cœlenterates.

Although some of the forms are very abundant, the number of species is not very great. I myself have observed three species in New Zealand, three species on the coast of Victoria, and five species in Port Jackson. Two of the latter are identical with the Victorian species. Of these nine species I was able to describe six; of the other three not sufficiently well preserved specimens were obtained by me. Of these six species only one had been previously

A

described, which makes it highly probable that only a very small percentage of the Australian species which occur in the places not visited by me have been hitherto described.

All the forms described by me are depicted in the plate, for the painting of which I am indebted to my mother.

This descriptive catalogue of the Australian species forms part of a series of similar works published by the Trustees of the Australian Museum, and I must here fulfil the pleasant duty of thanking them for the assistance afforded to me by them in continuing my researches into the Australian Cœlenterates. In particular am I indebted to the Curator of the Australian Museum, Mr. E. P. Ramsay, whose courtesy and judgment have made this work pleasant and efficient.

2. CLASSIFICATION OF THE CŒLENTERATES.

The Scyphomedusæ represent a very highly developed group of Cœlenterates. The largest forms are found among them, and the structure of the various organs of the adult is in some species more complicated than in any species of the other groups. The relationship of the Scyphomedusæ to the other equivalent groups can be expressed in the following manner :—

The Cœlenterata form the lowest type of the Metazoa, and can be divided into two sub-types, Porifera and Cnidaria, the former with and the latter without collared entodermal epithel cells in a part of the canal system.

The Porifera comprise the single class Spongiæ ; the Cnidaria are naturally divided into the Polypomedusæ, Polyps or their descendants, and Ctenaphoræ with external ribs of cilia, which are not descended from polypoid forms.

The Polypomedusæ again can be divided into two classes, Aphacellæ and Phacellatæ, according to whether they possess no entodermal gastral filaments, or whether such filaments are present. The second group comprises one class, the Ctenaphoræ.

The Aphacellæ are constituted of the two classes, Hydromedusæ and Syphonophora, whilst the Phacellatæ comprise the Scyphomedusæ and corals. I here adopt Ray Lankester's name Scyphomedusæ.

This classification is shown in tabular form on the following page, and is different from the systems commonly in use. In previous papers I have given my reasons for altering the classification used by Haeckel, Hincks, and others.

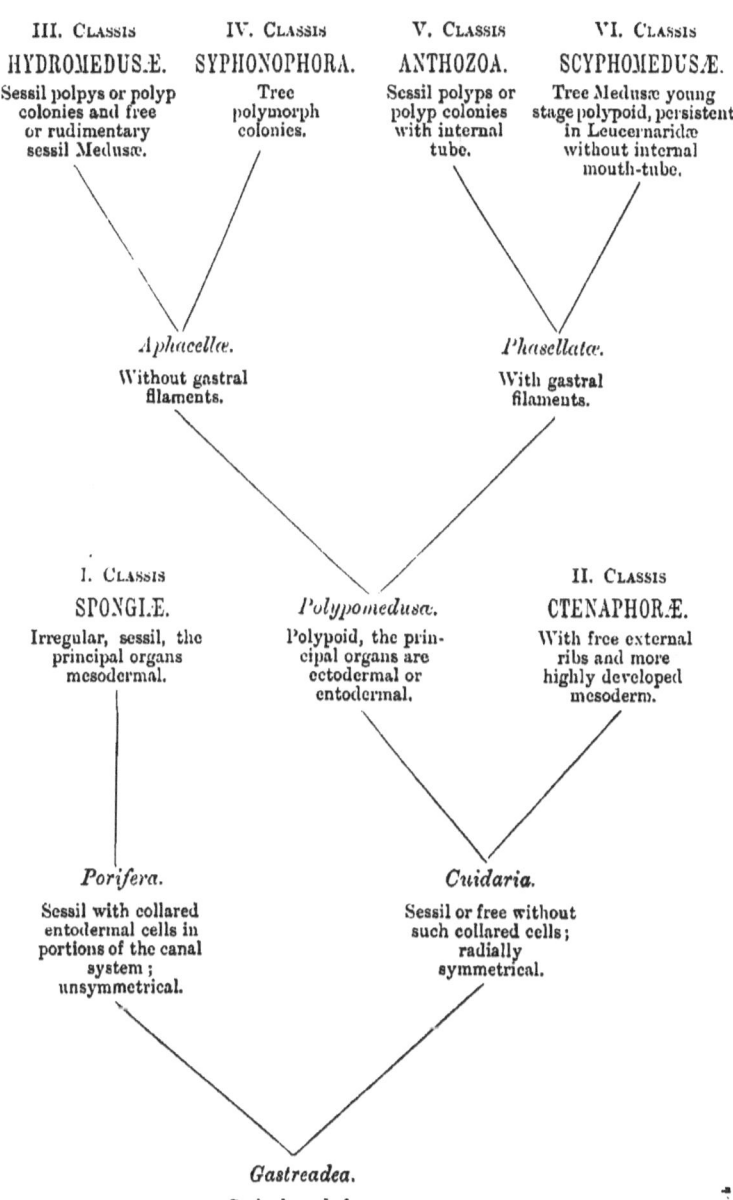

III. Classis
HYDROMEDUSÆ.
Sessil polpys or polyp colonies and free or rudimentary sessil Medusæ.

IV. Classis
SYPHONOPHORA.
Tree polymorph colonies.

V. Classis
ANTHOZOA.
Sessil polyps or polyp colonies with internal tube.

VI. Classis
SCYPHOMEDUSÆ.
Tree Medusæ young stage polypoid, persistent in Leucernaridæ without internal mouth-tube.

Aphacellæ.
Without gastral filaments.

Phasellatæ.
With gastral filaments.

I. Classis
SPONGIÆ.
Irregular, sessil, the principal organs mesodermal.

Polypomedusæ.
Polypoid, the principal organs are ectodermal or entodermal.

II. Classis
CTENAPHORÆ.
With free external ribs and more highly developed mesoderm.

Porifera.
Sessil with collared entodermal cells in portions of the canal system; unsymmetrical.

Cnidaria.
Sessil or free without such collared cells; radially symmetrical.

Gastreadea.
Sack-shaped, free-swimming, with simple ectoderm and entoderm.

3. The Life-history of the Scyphomedusæ.

The Jelly-fish are sexually differentiated, there are males and females. Hermaphrodites have hardly ever been observed. The sexual products are produced on the lower side of the umbrella, mostly in hernia-shaped extensions of the gastral wall. Here we find certain zones where either spherical ova in the female, or oval Spermasacs in the male, are found. In Chrysaora females small sack-shaped serata are met with in the sub-umbrella wall. In the female Pseudorhiza there are entodermal filaments, similar to the ordinary gastral filaments scattered all over the entodermal surface in the canals of the umbrella and arms, which appear filled with Spermasacs. Chrysaora females sometimes do not produce any male products. Pseudorhiza seems to be a true Hermaphrodite, no purely female specimens have been observed, nor do there seem to be any males. It appears that also the Linergedæ are Hermaphroditic in a similar manner as Chrysaora.

By the detachment of the ova they become free, and float about in the gastral cavity. Here they are fructified.

The ova may have a thick cell-wall with radial canals through it, a regular zona pellucida or their cell-wall is thin. When free they are not surrounded by a follicle which in some species, however, is present before birth.

The Spermatozoa are very slender, and have a sharp spearhead-shaped head and long tail. They are produced in great numbers in the Spermasacs by the bursting of which they become free. They then swim about in the gastral cavity of the male, and leave it after a little while.

Copulation, which might be supposed to occur has not been observed, as stated above, however, the fructification of the ova takes place *in the gastral cavity of the female.* The youngest stages are often found attached to the moutharms or in portions of the gastral cavity.

The gastrala is formed by invagination (Aurelia Chrysaora, &c.). Finally a ciliated larva is formed, which has an oval shape and consist of high cylindrical ectoderm cells and smaller cylinder cells in the interior. The one end, which, as the larva swims about, is anterior, is devoid of cnidablasts. There the ectoderm consists of high cylindrical gland-cells. These produce an adhesive slime by means of which the larva attaches itself to a suitable surface. In the ectoderm of the posterior end an abundance of cnidoblasts is met with.

The larvæ up to this stage are, in many species, carried about by the mother, and there are often peculiar arrangements found for their protection. In Cyanea annaskala the larvæ are attached to the entodermal side of the moutharms in such numbers that this side may attain a white colour. In Pseudorhiza aurosa peculiar sack-shaped appendages are found on the surface of the sub-umbrella, which at certain seasons appear filled with the larvæ. There are numerous other amusing arrangements in the different species, whilst in some, as for instance in our Crambessa mosaica, no larvæ have been observed in the body of the female medusæ.

At the time when the larva attaches itself the tentacles make their appearance. One is formed first, and becomes clearly visible before the second opposite one commences to grow. A pair of tentacles then grows out between the two primary ones, so that a polyp, with four tentacles and a mouth in the centre of them, results. Four longitudinal thickenings are then produced in the wall of the gastral cavity, which divide the latter into four separate chambers, analogous to the chambers in the gastral cavity of the Anthozoa. The number of tentacles increases in an irregular manner to sixteen. In most cases this is the limit of the number of tentacles. The Scyphystomæ of Crambessa mosaica, recently seen by me, possess thirty-two tentacles. This polyp, which is termed Scyphystoma, grows to a comparatively large size. It may reach, without changing its shape, a length of fifteen mm. These polyps multiply by budding to five or ten, which remain in connection by a tube-shaped stolon in such a manner that the gastral cavities are in free communication with each other. In some species the Scyphystoma does not multiply by budding, but remains solitary.

The whole of the ectoderm is charged with cnidoblasts. The tentacles are solid, and filled with a peculiar kind of entoderm, consisting of large cells, with thick walls. The mesoderm is developed only in the shape of a supporting lamella, and very thin. Only in the four longitudinal gastral ridges it is more massive. The entoderm consists throughout of uniform granular cells. Muscular fibres are found in the ectodermal lubepithel ; these belong to ectodermal "Neuromuscle cells."

When the Scyphystoma has attained its full size it commences to produce small medusæ by a process of strobilation. The distal portion of the Scyphystoma is divided into annular zones by transverse circular grooves, which make their appearance at first as insignificant depressed lines, which, however, soon increase in depth, while at the same time the inflated annular zones between them grow rapidly. Their growth commences at the upper distal end, so that we find the distal sections of the strobila much further developed than the proximal ones, and all in proportion to their distance from the base of attachment of the strobila.

The Scyphystoma, which by this process has been converted into a "Strabila," does not grow in size any more. The grooves become deeper, whilst the margin divides into a number of lobes. These are mostly eight in number. In Phyllorhyza I have observed sixteen. These lobes are again divided by a deep fissure into two secondary lobes, which are generally pointed. During these changes the groove has deepened until only a slender peduncle connects the young medusa, formed in this way, with the rest of the body of the strabila. This peduncle finally breaks off, and the young star-shaped medusa, which is termed an "Ephyra," swims away, and by gradual changes attains the structure of the adult jelly-fish.

The Ephyra is a small and thin star-shaped disc, measuring 1.5–3 mm. in diameter. The margin is regularly divided into eight or sixteen primary, and sixteen or thirty-two secondary lobes. The centre of the disc is occupied by the gastral cavity, which is lens-shaped, and from which a radial canal leads into each of the primary lobes. From the lower side a short quadrangular tube is suspended, which surrounds the mouth, situated at its base. The mouth is extended in the four primary radius, which being vertical on each other, give the mouth the peculiar shape of a cross. At the terminations of the eight or sixteen primary radial canals, and situated in the depressions between the secondary lobes, the marginal sense organs are found. These consist of a conic body, the smaller end pointing outward, and contain a number of crystalloid bodies in their distal entremity, the otoliths. They are hollow. The cavity is clothed with entoderm, and communicates with the primary radial canals.

Gastral filaments make their appearance in the Interradii on the lower side of the stomach.

Histologically the Ephyra is nearly as highly developed as the adult. The sexual organs are not yet formed, and the nerves, which radiate from the marginal sense organs, extend in the ectodermal epithelium instead of the sub-epithelium as in the adult.

It appears that not all the species of Scyphomedusæ undergo this change of generation. Pelagic species are developed from the invaginated gastrula direct in as much as the gastrula never becomes sessil, and produces a great amount of mesodermal gallert at an early stage, at the same time it attains a flattened shape, forms a mouth, and is directly converted into an Ephyra. Aurelia, although generally developing in the usual way by means of a change of generations, may under certain circumstances develop direct hypogenetically. Haeckel assumes that the Cubomedusæ, the development of which is unknown, may develop likewise in a hypogenetic manner without change of generation.

The Ephyra is converted into the adult Medusa by a process of simple Metamorphosis.

4. MORPHOLOGY OF THE SCYPHOMEDUSÆ.

The adult Scyphomedusæ are for the most part large, disc-shaped, free-swimming animals which are found in the sea. At breeding time many, particularly the Rhizostamidæ, swim up the rivers for some distance (Tajo, Parramatta, &c.), but never leave the salt or brackish water. No Scyphomedusæ have hitherto been found in fresh water. Certain forms, the Lucernaridæ, attach themselves at the aboral pole, and a similar sessile mode of life has been observed in certain Rhizostamidæ by *Keller* and others.

The Lucernaridæ resemble Scyphystomas very closely. No extensive disc-shaped umbrella filled with mesodermal gallert tissue is observed in them.

All the other Scyphomedusæ possess a large disc, which forms the bulk of the body above the stomach. This is termed the umbrella.

The Scyphomedusæ are radially symmetrical animals, and resemble in this particular the Echinodermata so closely that many of the older authors, and particularly *L. Agassiz*, combined the Scyphomedusæ, corals, &c., with the Echinoderms to form one group—the radiata. This classification can, however, not be retained in view of the great anatomical differences between these two, and it is not accepted at the present time.

The radii are not of equal morphological value, and it is easy to recognize that the organs are always in *fours*, forming a cross at the beginning (Scyphystoma), and that afterwards multiples of four are added to these. The mouth has always the shape of a cross, and the four radii which pass through the dilated corners of the mouth are known as the *primary radii*. They are vertical on each other. Between these, and cutting them at angles of 45°, there are four other radii to be distinguished, which pass through the sexual organs; these are the *secondary radii*. Between these eight again other radii of particular interest make their appearance; these are eight in number, and termed the *tertiary radii*. They cut the adjacent primaries and secundaries at the angle of 22° 30'.

The fundamental form of the Scyphomedusæ is that of a tetramer amphitect pyramid.

The body of an adult Scyphomedusæ consists of the following parts :—

1. A large circular disc or bell-shaped umbrella, which is very thick, and in the lower portion of which there extends a flat cavity, the stomach. The margin of the umbrella is serrated, flapped, or otherwise broken. The margin always is perfectly symmetrical. There are 4 x organs of sense—the Lucernaridæ are destitute of these *marginal bodies*—situated in indentures of the margin, and the margin consists of as many congruent autimers as there are marginal bodies. Each of these consists of two symmetric

halves. The plane of symmetry extends through the axis of the animal and the marginal body. The number of flaps in each such part is variable. It will be seen from this that if there are a marginal bodies ($a = 4x$) and b flaps in each symmetrical part of an autimer, that there will be a congruent autimers; $2a$, symmetrical parts; and $2ab$ marginal flaps on the umbrella.

The number x is generally 2—that is to say, that there are eight marginal bodies and autimers each with at least two, often four or more flaps. The The number x may, however, be 1 (Cubamedusæ), or greater than 2 (Collaspis, Bolyclonidæ, &c.).

2. In all Scyphomedusæ, with the exception of the Rhizostomæ, tentacles are developed, which grow out from the lower side of the umbrella near the margin; an exception is Aurelia. These are present also in the corresponding part of Lucernaridæ, and their number is very variable; they may be very numerous. The tentacles are solid or hollow, and in that case the canal which extends through them is in connection with the gastravascular system.

A kind of longitudinal chorda is developed along one side, and there are longitudinal striated muscular fibres in the ectodermal sub-epithelium.

The tentacles are studded with cuidablasts.

Their movements are very extraordinary, they can be contracted to one hundredth part of their length when expanded.

3. The marginal bodies referred to above, which are absent in the Lucernaridæ, are hollow cylindrical organs. The cavity is in communication with the canal system, and clothed with indifferent entodermal cells. Their outer surface is covered by differentiated ectoderm cells. The distal end is invariably occupied by a heap of otoliths. Further down pigment spots are frequently developed, which indicate the presence of an organ of sight. In Euhomedusæ, &c., this structure becomes very complicated, and five eyes with lenses, &c., can be distinguished on the marginal body (Charybdea, Claus). The narrow sensitive cells of the ectoderm are in connection with subepithelial ganglia cells. Often a conical depression is observed on the dorsal side of the umbrella just above the marginal body in which radiating folds and nerves (?) have been observed. Sometimes, also, complicated folds are formed at the base of the marginal body, and here sensitive cells and ganglia cells have been observed (Cyanea v. Lendenfeld). From the marginal bodies, nerves radiate towards the centre of the disc. These are situated on the lower side of the umbrella, just below the epithelial surface layer of cells in the adult (Cyanea v. Lendenfeld).

4. Systems of muscles are observed in the comparatively thin lamella which covers the stomach cavity. Also these are produced in the first

instance—phylogenetically—like the nervous elements from the ectodermal epithelium only, and situated in the sub-epithelium. The surface is folded so as to increase the extent and power of this muscular lamella, the projecting ribs are tough and elastic, and act like springs, inasmuch as they counteract the effect of the muscular contraction. The muscles are situated in very regular fields; there are circular and radial muscles. All are composed of band-shaped striated fibrils. By their contraction the umbrella is more curved, and when the contraction ceases the umbrella is again flattened out by the elasticity of the ribs mentioned above. By this movement of the umbrella the animal propels itself through the water. It is apparent that this apparatus for locomotion is not a very satisfactory one, and the movement, which appears most graceful but rather languid, results in a very slow motion of the animal through the water accordingly.

5. The stomach has more or less the shape of a very flat lens. From its margin canals arise, which sometimes form networks, and which supply the margin of the umbrella with nourishing material. The surface in which these canals are situated is marked by a lamella of entodermal cells—the entodermal lamella, which extends throughout the umbrella, dividing into two parts—a thick upper part without appendages, the ex-umbrella, and a thin oval part. In all the vital organs are situated the sub-umbrella. The stomach and canals are clothed with ciliated and gland cells of intodermal origin. By the continued vibration of the cilia, the contents of the gastro-vascular cavity are kept in incessant circulation. The Scyphomedusæ do not eat any large animals, but only soft and tender organisms like themselves.

6. The sub-umbral wall contains in four localities—in the secondary radii—appendages to the gastral cavity. In the walls of these the sexual products are matured. The outer shape of these genital organs is subject to great variations. They are generally very brilliantly coloured at the time when the sexual products are ripening in them. The latter have been described above. Often the four sexual organs of the secondary radii divide. Then there are eight which occupy the tertiary radii.

7. Generally the mouth is situated at the lower side in the middle of the stomach, and is cross-shaped. The margin of the mouth is produced in all the Scyphomedusæ, with the exception of the Rhizostomæ, where different structures are met with, to form lips. These may attain a very high degree of development, and grow out to form a long tube or large lamellons,* or membranes extensions, mouth-arms.

In the Rhizostomæ a vestibule is formed below the true stomach, and divided from it more or less completely. Only four columns, through each of which a canal passes, connect the vestibule stomach with the stomach proper

---

* *Sic.* ? lamellæ.

in the most differentiated genera. In others no such sub-gastral porticus is found, but only four deep depressions (Schirmhählen), dividing likewise the parochial disc from the umbrella. To the vestibule stomach eight large mouth-arms are appended, which are situated in the tertiary radii. These possess a deep groove leading to the vestibule stomach in the Chamoostomidæ. (Pseudorhiza, V. Lendenfeld.) In the true Rhizostomous medusæ the two sides of the groove form extensive folds, and coalesce in many places, so that numerous irregular and small apertures are formed instead of the simple one in other medusæ.

The mouth-arms are always, in whatever way they may be developed, clothed with entoderm on the inner, and with ectoderm on the outer side. In the margin the cnidoblasts are particularly numerous, and here also sensitive cells are found. The sub-epithelial layers contain muscular fibres, which extend in every direction, and enable the medusa to move its mouth-arms.

The Scyphomedusæ are present in all seas. The largest species (species of Cyanea) are found in the temperate zone.

Their geographical distribution depends to a great extent on the oceanic currents, but too little is known of the Australian species at present to allow of any conclusions regarding the distribution of the Australian species, to be described below.

THE AUSTRALIAN SPECIES.

In the following pages all Australian species hitherto discovered are described with full references. The classification is that of *Haeckel*, amended according to the results of the recent investigations of *Claus* and myself.

SCYPHOMEDUSÆ.

Ray Lancaster, 1877.

CŒLENTERATES without collared cells, medusoid with gastral filaments and without internal mouth tube.

I.—ORDO STAMOMEDUSÆ.

Haeckel, 1879.

Scyphomedusæ without sense organs, four or four pair of gonads in the sub-umbral wall, stomach with four large perradial pouches.

Familia LUCERNARIDÆ.

Johnston, 1847.

Margin of umbrella simple, undivided, without hollow arms or margin flaps, with simple tentacles. On the ex-umbrella a peduncle with which the medusa affixes herself.

Genus CRATEROLOPHUS. Clark, 1863.

Lucernaridæ with four mesagan pouches in the wall of the sub-umbrella, without marginal anchors or papillæ.

CRATEROLOPHUS MACROCYSTIS, R. von Lendenfeld.

Craterolophus macrocystis, R. v. Lendenfeld. The Scyphomedusæ of the Southern Hemisphere. Proceedings of the Linnean Society of New South Wales; volume ix, part 1, page 157.

The genus Craterolophus had only been found on the coast of Heligoland, where the hitherto single species Craterolophus Tethys (Clark) is very abundant, but nowhere else. Our species is accordingly the second. Unfortunately it is very rare, so that my description must be imperfect, as I only got two specimens, both of which were cut into sections forthwith.

Umbrella deep bell-shaped, expanded about half as broad as high (in Craterolophus Tethys broader than high). Stalk about $\frac{2}{3}$ of the height of the umbrella (in Craterolophus Tethys only $\frac{1}{4}-\frac{1}{3}$). Eight arms short at equal distances; every arm with a cluster of about thirty tentacles. The gonads are like those of Craterolophus Tethys, feathery.

Colour.—Dark olive green, fades in spirits.

Size.—Height of umbrella, 12 mm.; breadth, 6 mm.; stalk, 8 mm. high, and extended 3 mm. thick.

Locality.—East coast of New Zealand : Port Chalmers, Hutton, Lyttleton ; von Lendenfeld.

Ontogenesis.—Macrocystis.

II.—ORDO PERAMEDUSÆ.
Haeckel, 1879.

Scyphomedusæ with four marginal bodies which contain several eyes, four or twelve tentacles, eight or sixteen marginal flaps. Stomach surrounded by a large ring-shaped sinus, with four structures in the secondary radii. Four pair of frill-shaped gonads.

Familia PERICOLPIDÆ.
Haeckel, 1879.

Margin of umbrella with four tentacles and eight marginal flaps. Festoon canal, with sixteen pouches.

Genus PERICRYPTA. Haeckel, 1879.

Four pouches in the primary radii, with four continuous cavities. Four inter-radial tænials of the basal stomach ; they are high cavities covered along their whole length with two rows of gastral filaments.

PERICRYPTA GALEA, E. Haeckel.

Pericrypta galea, E. Haeckel. Das System der Medusen. Seite, 414.

Pericrypta galea, R. von Lendenfeld. The Scyphomedusæ of the Southern Hemisphere. Proceedings of the Linnean Society of New South Wales ; volume ix, part 1, page 166.

Umbrella high, helmet-shaped, 1½ as high as broad, divided into two equally high parts by a circular groove. The four perradial pedalia of the margin of the umbrella only slightly broader than the four interradial ones, slightly longer than the eight marginal laps. Four tentacles as long as height of umbrella. Mouth-tube cubic, with wide pouches with eight long adradial barbs.

Size.—Breadth, 30 mm. ; height, 4 mm.

Locality.—South Pacific Ocean : East coast of Australia, Schnehagen.

PERICRYPTA CAMPANA, E. Haeckel.

Pericrypta campana, E. Haeckel. Das System der Medusen. Seite 414.

Pericrypta campana, R. v. Lendenfeld. The Scyphomedusæ of the Southern Hemisphere. Proceedings of the Linnean Society of New South Wales ; volume ix, part 1, page 167.

Umbrella bell-shaped, a little higher than broad. Shallow circular groove. Inner part twice as high as marginal ring. The 4 perradial pedalia twice as broad as the 4 interradial ones, twice as long as the 8 stump margin laps. Four tentacles twice as long as the height of the umbrella. Mouth-tube quadrangular prismatic, half as long as the central stomach, without barbs.

Size.—Breadth of Umbrella, 24 mm.; height, 30 mm.

Locality.—South Pacific Ocean, near New Zealand, Weber.

Familia PERIPHYLLIDÆ.

Haeckel, 1879.

Peramedusæ with twelve tentacles and four marginal organs of sense, sixteen margin flaps. Exumbrella with sixteen pedalia, and circular muscle. On either side of the pedalia a pouch. Festoon canal consists of thirty-two flap-pouches.

Genus PERIPHYLLA. Steenstrup, 1837.

With four perradial pouches in the mouth-tube and four basal funnel cavities. The four interradial tacnials are hollow caves, along the whole length of which there are two rows of gastral filaments.

PERIPHYLLA MIRABILIS, E. Haeckel.

Periphylla mirabilis, E. Haeckel. Das System der Medusen. Seite 422.

Periphylla mirabilis, E. Haeckel. Die Tiefsee Medusen der "Challenger" Expedition. Seite 54. Tafel xviii, xxiii.

Periphylla mirabilis, R. v. Lendenfeld. The Scyphomedusæ of the Southern Hemisphere. Proceedings of the Linnean Society of New South Wales; volume ix, part 1, page 168.

Umbrella conic, ¼ higher than broad. Pedal zone higher than the lap zone. Both together ⅔ as high as the cone. Margin laps oval, distal wings triangular. Tentacle laps slightly projecting, less than the 8 laps of the marginal sense organs. Tentacles twice as long as height of umbrella, at the base ½ as broad as the margin laps. Mouth-tube cubic, ½ as high as umbrella, 8 adradial long barbs.

Colour.—In spirits, light violet, sub-umbral surface dark violet. Gonads orange.

Size.—Breadth of umbrella, 120 mm.; height, 160 mm.

Locality.—East coast of New Zealand. Latitude south 40° 28′; longitude east of Greenwich, 177° 43′. Depth, 1,100 fathoms. Station nr. 168, "Challenger." Wyville Thomson.

III.—ORDO CUBAMEDUSÆ.

Haeckel, 1879.

Scyphomedusæ with four perradial marginal bodies, which contain an acoustic club, with entodermal otolith sac and one or more eyes; four tentacles or tentacle clusters in the secondary radii. Gastral cavity with four wide square pouches in the primary radii. Gonads, four pair leaf-shaped bulges, which are fixed by their margin to the four septa in the secondary radii. They are developed from the sub-umbral entoderm of the pouches of the stomach, and they project free into the cavity.

Familia CHORYBDEIDÆ.

Gegenbaur, 1856.

Cubamedusæ with four simple tentacles in the secondary radii, and four marginal bodies in the primary radii, without any marginal flaps in the velarium, but with eight marginal pouches, without arms in the four radial pouches.

Genus PROCHARYBDIS. Haeckel, 1879.

Chorybdeidæ with four simple interradial tentacles with pedalia, simple velarium without velar canals and without trenula.

PROCHARYBDIS FLAGELLATA, E. Haeckel.

Procharybdis flagellata, E. Haeckel.   Das System der Medusen.   Seite 438.

Procharybdis flagellata, R. v. Lendenfeld.   The Scyphomedusæ of the Southern Hemisphere.   Proceedings of the Linnean Society of New South Wales; volume ix, part 2, page 273.

Marsupialis flagellata.   Lesson acalèphes, page 278.

Umbrella conic stubbed above. The height twice the width; four side surfaces strongly curved. The four interradial sides rounded and a little projecting. Gastral cavity? Margin of umbrella continuous with eight laps. Velarium simple, narrow, continuous. The distance of the niches of the sense organs to the margin of the umbrella is about half the distance of the pedal base. Four pedalia lancet-shaped and half as long as the height of the umbrella, with very narrow wings. Tentacles several times as long as the height of the umbrella.

Size.—Breadth of umbrella, 30 mm.; height of umbrella, 40 mm.; ontogenesis unknown.

Locality.—North coast of Australia: Torres' Straits, Weber. (New Guinea, Lesson?)

IV.—Ordo DISCAMEDUSÆ.
Haeckel, 1866.

Scyphomedusæ with 8—16 or more marginal bodies (always at least in the primary and secondary radii) ; in each marginal body an acoustic vesicle with entodermal otolith pouch, and often at the same time an eye. Marginal flaps, always 8 pairs primary (ephyraflaps), and besides often numerous accessory (velarflaps). Tentacles present or absent. Gastral cavity surrounded by a circle of radial processes (8—12—16 or more), sometimes broad radial pouches, sometimes narrow radial canals. Gonads 4, in the secondary radii, representing folded bulges in the sub-umbrella of the sides of the gastral cavity, from the entoderm of which they are developed (they rarely divide to form eight bulges situated in the tertiary radii). Umbrella flattened orbicular.

I.—Sub-ordo CARMASTOMÆ.
Haeckel, 1879.

Discamedusæ with simple quadrangular mouth tube, without mouth-arms, with simple central mouth and with short solid tentacles.

Familia EPHYRIDÆ.
Haeckel, 1879.

With broad radial pouches, without terminal branch canals, with simple four cornered manubrium, without mouth-arms, with simple central mouth. Mostly 16 broad radial pouches (8 ocular and 8 tentacular), rarely 16—32. With these alternating as many short, solid tentacles. Mostly 16 (rarely 32—64) marginal flaps. Simple flap pouches present or absent. Branched flap canals absent. Four or four pair of gonads in the secondary radii situated in the sub-umbral wall.

Genus EPHYRA. Haeckel, 1879.

Ephyridæ with 8 marginal bodies and 8 tentacles with 16 marginal flaps, without flap pouches, and four horse-shoe-shaped gonads.

EPHYRA PROMETOR, E. Haeckel.

Ephyra prometor, E. Haeckel. Das System der Medusen. Seite 482, tafel xxvii, figure 1, 2.

Ephyra prometor, R. v. Lendenfeld. The Scyphomedusæ of the Southern Hemisphere.

Umbrella flat, bell-shaped, 1½ times as broad as high. Marginal flaps oval, about as long as broad, and ⅓ as long as the radius of the umbrella. Tentacles pointed, twice as long as the marginal flaps. Four simple interradial gastral filaments. Four gonads, horse-shoe-shaped, without flaps, and smooth.

Size.—Breadth of umbrella, 8 mm. ; height of umbrella, 6 mm. Onto-genesis unknown.

Locality.—Coast of Australia, Weber.

II.—Sub-ordo SEMOSTOMÆ.

L. Agassiz, 1862.

Discomedusæ, with 4 large folded mouth-arms, with simple central mouth, and long, hollow tentacles.

Familia PELAGIDÆ.

Gegenbaur, 1856.

With a simple, cross-shaped mouth, and 4 folded mouth-arms in the primary tentacles. With simple, broad radial pouches; without branched distal canals; without ring canal; 8 marginal bodies; 16—32 or more marginal flaps.

Genus PELAGIA.

Péron et Lesueur, 1809.

Pelagidæ with 8 tentacles in the tertiary radii (alternating with the 8 marginal bodies), and with 16 marginal flaps.

PELAGIA PANOPYRA, Péron et Lesueur.

Pelagia panopyra, Péron et Lesueur.   Tableaux des Méduses, &c.; page 349, n. 64.

Pelagia panopyra, E. Haeckel.   Das System der Medusen.   Seite 509.

Pelagia panopyra, F. Eschscholz.   System der Acalephen ; page 73, tafel vi, figure 2.

Pelagia panopyra, R. Lesson (p.p. !).   Centurie Zool ; page 192, plate 62, figure 2.

Pelagia panopyra, Brandt.   Mémoire Acad. Petersb. ; tome IV, page 382, tafel xiv, figure 1, tafel xiv, A, figures 1–5.

Pelagia panopyra, L. Agassiz.   Contributions to the Natural History of the United States ; iv, page 164.

Pelagia tuberculosa, Couthony.   L. Agassiz.   Contributions ; iv, page 164.

Pelagia Labiche, F. Eschscholz (?).   System der Acalephen; page 78.

Pelagia Labiche, H. de Blainville (?).   Actinologie ; page 302, plate 40, figure 3.

Pelagia Labiche, L. Agassiz (?).   Contributions to the Natural History of the United States ; iv, page 165.

Medusa panopyra, Péron et Lesueur.   Voyage aux Terres Australes ; plate, 31, figure 2.

Dianæa panopyra, T. de Lamarck.   Histoire naturelle des Animaux sans Vertèbres ; tome II., page 507.

Cyanea Labiche, Quoy et Gaimard. Voyage de l'Uranie, etc. Zoologie, page 571, plate 84, figur 1.

Pelogia panopyra, R. v. Lendenfeld. The Scyphomedusæ of the Southern Hemisphere. Proceedings of the Linnean Society of New South Wales ; volume ix., part 2, page 266.

Umbrella semi-spherical, flattened above, nearly twice as broad as high. Nettlewarts of the ex-umbrella, small and scattered, elongated ; most dense on the margin of the umbrella. Marginal flaps nearly quadratic, slightly crenated on the distal margin. Manubrium long and narrow, nearly twice as long as the radius of the umbrella ; three times as long as broad. Mouth-arms long and narrow, one and a half times as long as the manubrium ; about three times as long as the radius of the umbrella ; its membranous border is twice as broad as the thin cylindrical middle-rip at its base.

Colour.—Variable, generally pale-rose or violet ; mouth-arms more violet ; gonads more purple ; nettle-warts violet.

Size.—Breadth of umbrella, 50 mm. ; height of umbrella, 33 mm.

Ontogenesis.—Unknown.

Locality.—Tropic Zone of the Pacific Ocean, from Australia to Peru, Péron, Eschscholtz, Lesson, von Mertens, Couthony, &c.

Familia CYANIDÆ.

L. Agassiz, 1862.

Discamedusæ, with a simple cross-shaped mouth, surrounded by four folded mouth-arms. Gastral cavity, with 16 or 32 broad radial pouches, and branched cœcal flap-canals, without a ring canal (8 or 16 marginal bodies), and 8 or more hollow tentacles.

Genus STEMOPTYDIA. L. Agassiz, 1862.

Cyanidæ with 8 marginal bodies and 40 tentacles, 5 on each tentacle flap. Umbrella with 8 main and 16–32 secondary flaps.

STENOPTYCHA ROSEA, L. Agassiz.

Stenoptycha rosea, L. Agassiz. 1862. Monogr., Acal., Contrib., iv, page 162.

Cyanea rosea, Quoy et Gaimard. 1827. Voyage de l'Uranie, etc. Zoologie, page 570, plate 85, figure 1, 2.

Stenoptycha rosea, E. Haeckel. Das System der Medusen. Seite 525.

Stenoptycha rosea, R. v. Lendenfeld. The Scyphomedusæ of the Southern Hemisphere. Proceedings of the Linnean Society of New South Wales ; volume ix, part 2, page 272.

B

Umbrella slightly vaulted, semi-spherical two or three times as broad as high. Ex-umbrella set with warts, covered with pointed elevations. Mouth-arms tender, richly folded curtains, about as long as the radius of the umbrella. The margin of the umbrella has sixteen small incisions of which the eight ocular incisions are deeper than the eight tentacular ones, sixteen flaps quadrangular, truncate. There are five very large tentacles on the ventral side of each main flap, they are ten times as long as the diameter of the umbrella.

Colour.—Pink, margin of umbrella and tentacles, darker.

Size.—Breadth of umbrella, 200 mm. ; height of umbrella, 100 mm.; length of tentacles, nearly 2 metres.

Locality.—Port Jackson, Sydney, Quoy and Gaimard.

Genus CYANEA   Péron et Lesueur, 1809.

Cyanidæ with eight marginal bodies and numerous tentacles which form eight bundles of in the tertiary radii on the sub-umbrella. There are several rows of tentacles, one behind the other in each bundle. Margin of umbrella with eight main flaps and 16–32 secondary flaps.

CYANEA ANNASKALA, R. von Lendenfeld.

Cyanea annaskala, R. v. Lendenfeld.   Ueber coelenteraten der Südsee.   1. Mittheilung.   Freitschrift für wissenschaft liche Zoologie.   Band xxxvii. Seite 466, Tafel xvii—xxiv.

Cyanea annaskala, R. v. Lendenfeld.   The Scyphomedusæ of the Southern Hemisphere.   Proceedings of the Linnean Society of New South Wales ; volume ix, part 2, page 275.

Cyanea annaskala, R. v. Lendenfeld.   Local colour.   Varieties of Scypho-medusæ   Proceeding of the Linnean Society of New South Wales ; volume ix, part 2, page 925.

The umbrella flaps are rounded and not broader at the end than at the base.   Every main flap (Ephyra-arm) consists of four flaps, two smaller ocular flaps and two larger flaps at the sides.   The umbrella is 5—7 times as broad as high, depressed, with a few protruding nettle-warts in the centre of the ex-umbrella.

Ontogenesis.—The embryos hang on to the mouth-arms until they are nearly matured to young Scyphystomæ, and then affix themselves to bodies in the water, producing a long stalk with a chitinous perisark and eight arms.   The Ephyra passes into the adult by a complicated metamorphosis.   The umbrella flaps are produced by fission.

Colour.—Umbrella and tentacles colourless. Entoderm of the gastral cavity brown; mouth-arms white, with purple margin, or purple throughout. Genital organs in the male rose-coloured, and in the female orange-yellow.

Size.—70—500 mm. diameter of the umbrella; length of tentacles, 300 mm.

I. CYANEA ANNASKALA VAR PURPEREA.

A small variety not exceeding 150—200 mm. in diameter, with mouth-arms, which are of a uniform purple colour throughout.

Locality.—Port Phillip, v. Lendenfeld. Abundant in January, February and March.

II.—CYANEA ANNASKALA VAR MARGINATA.

A large variety, attaining a diameter of 400 mm. in the umbrella. The mouth-arms are colourless, only on the margin there is an intensely purple zone a few millimetres in width.

Locality.—Port Jackson, v. Lendenfeld. From November to March.

Familia AURELIDÆ.
Claus, 1883.

Lemostomæ with narrow canals which are joined by a ring-canal, with very small ephyra-flaps (ocular flaps), with numerous short hollow tentacles, which arise from the dorsal side of the umbrella on the 8 intermediate flaps in the tertiary radii. With cavities (schirmhahlen) for the sexual organs.

Genus AURELIA. Péron et Lesueur, 1809.

Mouth-arms simple.

AURELIA CLAUSA. Lesson.

Aurelia clausa, R. Lesson. Voyage de la Coquille. Zool., p. 119.

Aurelia clausa, E. Haeckel. Das system der Medusen. Seite 558.

Aurelia clausa, L. Agassiz. Monograph. Contributions to the Natural History of the United States of America; vol. iv., page 160.

Claustra pissimbogue, R. Lesson. Acalèphes, page 78.

Ocyroe lineolata, Péron et Lesueur. Tableaux des Méduses, page 355.

Cassiopea lineolata, T. de Lamarck. Système des Animaux sans vertèbres. Tome ii, page 511.

Aurelia clausa, R. v. Lendenfeld. The Scyphomedusæ of the Southern Hemisphere. Proceedings of the Linnean Society of New South Wales; volume ix, part 2, p. 279.

Umbrella semi-spherical, twice as broad as high, 16 velar flaps of the umbrella margin protruding. Mouth-arms narrow, thin and curled, they coalesce at the base, with a large oval pointed lip-like thickening. The four labial thickenings can close (?) the entrance to the sub-genital pouches.

Colour.—Ovaries, canals, and tentacles rose coloured.

Size.—Breadth of umbrella, 80–100 mm. ; height, 40–50 mm.

Locality.—South Pacific Ocean, Port Praslin, New Ireland, Lesson ; New Zealand, Australia (?) Péron.

AURELIA CÆRULEA, R. von Lendenfeld.

Aurelia cærulea, R. v. Lendenfeld. The Scyphomedusæ of the Southern Hemisphere. Proceedings of the Linnean Society of New South Wales ; volume ix, part 2, page 280.

The umbrella is very flat, about 5 times as broad as high, with 16 small ocular and 16 velar flaps, divided from the former by deep incisions, but from each other only by a shallow groove. The mouth-arms are broad and a little longer than the margin of the umbrella. They are rounded at the end. The centrifugal branches of the canal system divide at larger angles than in other species, so that the ramification has by no means a slender appearance. There are very few anastomoses only.

It differs from Aurelia aurita (Lamarck) by its broad mouth-arms, the margins of which are not curled, and by the stubby appearance of the canal system. In these respects Aurelia colpota (Brandt) resembles Aurelia cærulea. The colour of our species is similar to that of Aurelia aurita, but fainter, and always decidedly blue. Aurelia colorata also differs from our species in colour. Aurelia flavidula (Péron et Lesueur) possess flaps at the base of the mouth-arms, and has a yellow tinge. Aurelia marginalis (L. Agassiz) has much smaller mouth-arms. Aurelia hyalina (Brandt), again, possesses lancet-shaped mouth-arms. Aurelia labiata (Chamisso et Eysenhardt) possesses 16 velar flaps like our species, but these are divided from each other by deep incisions. In Aurelia clausa (Lesson) the mouth-arms are narrow and curled, and the colour is red. In Aurelia limbata (Brandt) there also are 16 very distinct velar flaps, and the tentacles are brown.

Similar in appearance to one or other of these species, it still appears advisable to distinguish Aurelia cærulea from these. The main feature of our species are the broad and smooth mouth-arms.

Colour.—Transparent and blue. Only the hoof-shaped gonads reflect the light, and appear white or light rose-coloured, as in Aurelia aurita.

Size.—Diameter of umbrella, 110 mm. ; height, 20–30 mm.

Ontogenesis.—I have obtained a single larva measuring 9 mm. This being similar to corresponding stages in European Aurelias, we may suppose that the Ontogenesis of our species is similar to that of Aurelia aurita.

Locality.—Port Jackson, von Lendenfeld.

III.—Sub-ordo RHIZOSTOMÆ.

Cuvier.

Discomedusæ with eight large root-shaped simple or branched mouth-arms in the tertiary radii, with numerous suction-mouths on, or a deep furrow in, each mouth-arm, without tentacles.

Familia CHAMIOSTOMIDÆ.

Von Lendenfeld, 1882.

Rhizostomæ with a continuous sub-genital porticus. The mouth-arms which originate from the brachial disc are pinnately branched, and possess on the ventral side a deep groove, bridged over in one locality only to form a filament, by the local coalescing of the membranous lateral extensions. Scattered serata throughout the gastro-vascular system.

Genus PSEUDORHIZA. Von Lendenfeld, 1882.

With one large filament at the bifurcation of one or all eight arms. The canal system consists of 16 radial canals and a ring canal. Centrifugally from the ring canal we find an anastamosing network of canals, whilst centrifugally there are between two adjacent main radial canals, ten canals running from the ring canal inwards radially.

PSEUDORHIZA AUROSA. R. von Lendenfeld.

Pseudorhiza aurosa, R. v. Lendenfeld. Zoologischer Anzeiger, Nr. 116; band v, Seite 380.

Pseudorhiza aurosa, R. v. Lendenfeld. The Scyphomedusæ of the Southern Hemisphere. Proceeding of the Linnean Society of New South Wales; Volume ix, part 2, page 293.

Umbrella three or four times as broad as high, with a reticulate figure on the dorsal side. In every octant between the two long and narrow ocular flaps there are six velar flaps, each of which consists of three secondary flaps. Arms about as long as the diameter of the umbrella.

Colour.—Umbrella colourless, with a violet reticulate figure covering the ex-umbrella. Entoderm of gastral cavity, brown; upper part of mouth-arm-grooves, rose coloured; arm, colourless and transparent; frills along the margin of the grooves, and distal end of long filament, richly violet.

Size.—Diameter of umbrella, 400 mm.; thickness of gallert of umbrella, 30 mm.; brachial disc, 220 mm. in diameter, and 25 mm. in thickness.

Ontogenesis.—The embryos are carried about in pouches suspended in great number from the radial canals, which run centripetally from the ring canal.

They remain there till they are fit to turn into Scyphystomas, then escape and affix themselves with the aboral pole to any free submerged surface. The Scyphystoma does not differ from other Scyphystomas.

Locality.—Port Phillip, von Lendenfeld; Glenelg, Haacke.

PSEUDORHIZA HAECKELI, Haacke.

Pseudorhiza Haeckeli, W. Haacke.   Pseudorhiza Haeckeli spec. nov. der cnd-spross des discoundusenstammes biologisches centralblatt. Bond 4, Nr. 10. Seite 291.

Similar in every respect to the above species, but with only one single fila-ment on one arm. It seems to me not impossible that my species Pseudorhiza aurosa may be a younger stage of this one. There seem, however, to be dif-ferences between them also apart from the difference in the development of the filaments. No diagnosis is given by Haacke in the above loc. Dr. Haacke was so kind as to furnish me with a drawing of his species, and it appears that the solitary filament of Pseudorhiza Haeckeli is much thicker, one about 3 cm, than the filaments in Pseudorhiza aurosa, which only measures ½ cm in thickness.

Ontogenesis—Unknown.

Locality.—St. Vincent Gulf, Haacke.

Familia ARCHIRHIZIDÆ.

Claus, 1883.

Rhizostomæ of small size with eight unbranched mouth-arms, without ter-minal branches, with simple canal system. The radial canals are joined by a ring canal, and form only very few ramifications.

Genus ARCHIRHIZA. Haeckel, 1879.

Mouth-arms without vesicles or filaments, with sixteen radial canals, ring canal, and eight marginal bodies.

With four distinct sub-genital pouches.

ARCHIRHIZA PRIMORDIALIS. E. Haeckel.

Archirhiza primordialis, E. Haeckel.   Das System der Medusen.   Seite 565. Tafel xxxvi. Fig. 1, 2.

Archirhiza primordialis, R. v. Lendenfeld.   The Scyphomedusæ of the Southern Hemisphere.   Proceedings of the Linnean Society of New South Wales; volume ix., part 2, page 282.

Umbrella slightly vaulted, hat-shaped or semi-spherical, 2—3 times as broad as high. Umbrella margin with 48 marginal flaps. In every octant four large pointed velar flaps, and two small ocular flaps. Mouth-arms about as long as the umbrella radius, with simple zigzag-shaped suction crisp.

Size.—Diameter of umbrella, 40 mm. ; height, 20 mm.

Locality.—Bass Straits, Smith.

ARCHIRHIZA AUROSA. E. HAECKEL.

Archirhiza aurosa, E. Haeckel. Das System der Medusen. Seite, 645.

Archirhiza aurosa, R. v. Lendenfeld. The Scyphomedusæ of New South Wales. Proceedings of the Linnean Society of New South Wales; volume ix., part 2, page 582.

Umbrella flat, disc-shaped, margin, with 80 flaps, in every octant 8 oval pointed velar flaps between two broad triangular ocular flaps ; mouth-arms one and a-half times as long as the umbrella radius, conic, simple, with simple suction crisp.

Size.—Breadth of umbrella, 50 mm. ; height, 20 mm.

Locality.—New Zealand, Weber.

Genus HAPLORHIZA. Haeckel, 1879.

The mouth-arms do not coalesce with their lateral walls, and are unbranched. Sub-genital pouch continuous, dividing the brachial disc completely from the umbrella.

HOPLORHIZA PUNCTATA. E. Haeckel.

Hoplorhiza punctata. E. Haeckel. Das System der Medusen. Seite 604.

Hoplorhiza punctata. R. v. Lendenfeld. The Scyphomedusæ of the Southern Hemisphere. Proceedings of the Linnean Society of New South Wales ; volume ix, part 2, page 294.

Umbrella slightly vaulted with 176 marginal flaps (in each octant 10 pairs of narrow, round velar flaps, between 2 small recurved ocular flaps), 8 arms quite simple, cylindrical, half as long as the radius of the umbrella. Sub-genital ostia three times as broad as the pillars.

Colour.—Ex-umbrella dark violet-brown, equally speckled with white spots.

Size.—Breadth of umbrella, 40 mm. ; height of umbrella, 20 mm.

Ontogenesis.—Unknown.

Locality.—Coast of North Australia, Arnheim's Land, J. M. Elsey.

HOPLORHIZA SIMPLEX. E. Haeckel.

Hoplorhiza simplex. E. Haeckel. Das System der Medusen. Seite 604

Hoplorhiza simplex. R. v. Lendenfeld. The Scyphomeduse of the Southern Hemisphere. Proceedings of the Linnean Society of New South Wales ; volume ix, part 2, page 293.

Umbrella flat orbicular with 48 marginal flaps (in each octant 4 large square velar flaps between 2 tongue-shaped projecting ocular flaps) ; 8 arms quite simple, cylindrical, as long as the radius of the umbrella. Subgenital ostia about as broad as the distance between them.

Size.—Breadth of umbrella, 50 mm. ; height of umbrella, 20 mm.

Ontogenesis.—Unknown.

Locality.—South Australia, Bass Straits. Museum, Godeffroy.

Genus CANNORHIZA. Haeckel, 1879.

With eight simple mouth-arms, which are connected with their lateral margins, and form together an octagonal tube, with one opening underneath. Sub-genital porticus continuous. Parochial disc completely separated from the umbrella.

CANNORHIZA CONNEXA. E. Haeckel.

Cannorhiza connexa, E. Haeckel. Das System der Medusen. Seite 605.

Cannorhiza connexa, R. v. Lendenfeld. The Scyphomeduse of the Southern Hemisphere. Proceedings of the Linnean Society of New South Wales ; volume ix, part 2, page 294.

Umbrella flat, orbicular, with eighty marginal flaps (in each octant eight narrow rectangular velar flaps between two oval-pointed ocular flaps). Brachial tube formed by the lateral concrescence of eight cylindrical simple mouth-arms, a little longer than the radius of the umbrella. Above, half as broad as long, below, the same breadth. The four wide sub-genital ostia about as broad as the strong and free pillars between them. Brachial disc about as broad as the radius of the umbrella.

Size.—Breadth of umbrella, 80 mm. ; height of umbrella, 30 mm.

Ontogenesis.—Unknown.

Locality.—South Pacific Ocean, near New Zealand, Smith.

Familia CASSIOPÆIDÆ.
Claus, 1883.

Rhizostomæ with broad flat brachial disc. The eight mouth-arms are long, and possess club-shaped appendages, which are vesicles. The upper-arm continued to form the main branch of the lower arm. Canal nets very dense and with fine meshes, mostly with numerous radial canals.

Genus TOREUMA. Haeckel, 1879.

With eight primate or trichotome mouth-arms. The upper arm continued beyond the first ramification to form the main branch of the under arm in the tertiary radius. Numerous club-shaped vesicles between the arm crisps 16 radial canals, 8 marginal bodies. Four distinct sub-genital pouches.

TOREUMA THEOPHILA, E. Haeckel.

Toreuma theophila, E. Haeckel. Das System der Medusen. Seite 566.

Cassiopœa dieuphida, Péron et Lesueur. Tableaux des Méduses, page 356.

Cassiopœa theophila, T. de Lamarck. Histoire Naturelle des Animaux sans Vertèbres, tome ii, page 511.

Rhizostoma theophila, Fr. Eschscholtz. System der Acalephen. Seite 53.

Polydoma theophila, L. Agassiz. Monograph of the Acalephes. Contributions to the Natural History of the United States; volume iv., page 159.

Toreuma theophila, R. v. Lendenfeld. The Scyphomedusæ of the Southern Hemisphere. Proceedings of the Linnean Society of New South Wales; volume ix., part 2, page 283.

Umbrella semi-spherical 2—3 times as broad as high. Margin with 96 short coalescing flaps, in each octant 10 velar flaps between two very small velar flaps. Ex-umbrella roughly granulated, studded with warts, with small oval white spots on the marginal flaps. Eight arms about as long as the umbrella radius. With three or four pair of broad and flat main branches. Between the clusters of crisps numerous small and 10 or 20 large club-shaped vesicles. The latter two or three times as long as the breadth of the main branches.

Colour.—Umbrella, brownish-red, with white spots on the margin flaps. Gonads and vesicles, white.

Size.—Diameter of umbrella, 60—80 mm. ; height, 20—30 mm.

Locality.—Nord-west coast of Australia, de Witt's Land, Péron.

Genus VERSURA. Haeckel, 1879.

With 8 pinnate or trichotome mouth-arms, mouth-cross with perradial forked suctorial crisps, without central frils. Sub-genital pouch continuous. Brachial disc derided from the umbrella.

VERSURA VESICATA, E. Haeckel.

Versura vesicata, E. Haeckel. Das system der medusen. Seite 645.

Versura vesicata, R. v. Lendenfeld. The Scyphomedusæ of the Southern Hemisphere. Proceedings of the Linnean Society of New South Wales; volume ix, part 2, page 296.

Umbrella flat scutiform, with eight deep ocular incisions, 208 coalescing marginal flaps (in each octant 24 narrow rectangular velar flaps between two rudimentary ocular flaps). Sub-genital ostia half as broad as the pillar between them. Arms about as long as the radius of the umbrella, doubly pinnate, flat spread out as long as broad.

Locality : Australia (North-west coast)? Weber.

Familia CEPHEIDÆ.
Claus, 1883.

Rhizostomæ with broad flat brachial disc. The eight mouth-arms form by branching dichotomously two terminal branches. They are studded with long nettle-clubs or filaments. Canal nets very dense with small meshes and generally with numerous radial canals.

Genus CEPHEA. Péron et Lesueur, 1809.

With eight bifurcate mouth-arms. The branches are simple and not dichotomous. The upper arm does not extend beyond the bifurcation. Four distinct Euligneital pouches.

CEPHEA FUSCA, Péron et Lesueur.

Cephea fusca, Péron et Lesueur. Tableaux des Méduses, page 361.

Cephea fusca, E. Haeckel. Das System der Medusen. Seite 575.

Cephea fusca, Eschscholtz. System der Acalephen. Seite 57.

Cassiopea fusca, Dussumier. Musée du Jardin des plantes, No. 111.

Polyrhiza fusca, L. Agassiz. Contributions to the Natural History of the United States, iv, page 156.

Cephea fusca, R. v. Lendenfeld. The Scyphomedusæ of the Southern Hemisphere. Proceedings of the Linnean Society of New South Wales; volume ix, part 2, page 286.

Umbrella cap-shaped, central dome flat vaulted, covered with sixteen or twenty large conic excrescences. Divided from the thin margin of the umbrella by a deep circular furrow. In every octant eight flaps which appear as secondary flaps of one large primary one which lies between the two ocular flaps. The two flat branches of the under arm three times as long as the simple upper arm. They do not reach to the margin of the umbrella. Filaments numerous, decreasing in size centrifugally. The longest as long as the umbrella radius.

Colour.—Umbrella, dark brown. Ex-umbrella, with eight white radial stripes. Arms, yellowish ; filaments, white.

Size.—Breadth of umbrella, 150 mm. ; height, 30 mm.

Locality.—North-west coast of Australia : de Witt's Land, Péron et Lesueur ; Malabar, Dussumier.

Genus PHYLLORHIZA. L. Agassiz, 1862.

With eight pinnate mouth-arms. The pinnæ rudimentary and connected by a membrane, so that the whole arm attains an extended leaf-shaped appearance. Continuous sub-genital porticus. Brachial disc completely divided from the umbrella.

PHYLLORHIZA PUNCTATA, R. von. Lendenfeld.

Phillorhiza punctata, R. von. Lendenfeld. The Scyphomedusæ of the Southern Hemisphere. Proceedings of the Linnean Society of New South Wales; volume ix, part 2, page 296.

The umbrella is nearly semi-spherical, and about half as high as broad. The umbrella margin bears in every octant two sickle-shaped ocular flaps, four simple and four double flaps, all of which taper centrifugally and are truncate. These flaps consist of thick gallert, and are connected by thin membranes. Radial furrows extend centripetally from the fissions towards the centre of the disc. The sub-genital ostia are large and oval, more than twice as broad as the pillars of the brachial disc. The brachial disc is thick and a little more than half as broad as the umbrella, octagonal, with a canal system of its own, and thickly set on its ventral side with filaments, to which the young gastrulæ-embryos adhere. The mouth-arms are $\frac{2}{3}$ as long as the diameter of the umbrella.

Colour.—Brachial disc, stem and branches of mouth arms, and also the ground substance of the umbrella, colourless. In the gallert of the umbrella, close to the surface, are groups of "yellow cells," zooxanthellæ, which give the whole a brown colour. Towards the exumbral surface there are clouds of strongly refracting minute bodies in the gallert, which give our animal the spotted appearance from which I have derived the specific name. The suctorial crisps are brown. The long-stalked suction naps (Peitschen filamente) are colourless and transparent.

Size.—Breadth of umbrella, 500 mm. ; thickness of ex-umbrella, 50 mm. ; mouth-arms, 300 mm. ; filaments 200 mm. long, and at the base 10 mm. thick.

Ontogenesis.—The young embryos adhere to the mother's filaments until they have nearly attained the scyphystoma stage. The ephyra is brown spotted. The young medusa goes through a complicated metamorphosis, passing through stages with 24 and 16 marginal bodies.

Locality.—Port Jackson, Von Lendenfeld.

Familia CRAMBESSIDÆ.

V. Lendenfeld.

With broad styliform elongate brachial disc. The mouth-arms terminate in a long under-arm which consists of three wing-shaped parts, and possess dorsal crisps. Often with club-shaped gallert-knots, canal-system with 8 radial canals and a distal close reticulation. Continuous sub-genital porticus and free brachial disc.

Genus CRAMBESSA.

Haeckel, 1879.

Crambessidæ, with free, strong upper-arm, and strong triangular pyramidal under-arm, the 3 broad wings of which possess suctorial crisps without terminal knots or cripless appendages.

CRAMBESSA PALMIPES, E. Haeckel.

Crambessa palmipes, E. Haeckel. Das System der Medusen. Seite 620.

Crambessa palmipes, R. v. Lendenfeld. The Scyphomedusæ of the Southern Hemisphere. Proceedings of the Linnean Society of New South Wales; volume ix, part 2, page 299.

Umbrella semi-spherical, twice as broad as high, with 64 marginal flaps. In each octant 6 square truncate velar-flaps between 2 projecting oval ocular-flaps, which are half as long as the former. Ex-umbrella finally granulated; arms a little shorter than the radius of the umbrella; under-arm triangular-pyramidal, pointed, surrounded by suctorial crisps, twice as long as the short and cylindrical upper-arms, which are connected with a thin membrane.

Size.—Breadth of umbrella, 40 mm. ; height of umbrella, 20 mm.

Ontogenesis.—Unknown.

Locality.—North-coast of Australia, Koch. Museum, Godeffroy.

CRAMBESSA MOSAICA, E. Haeckel.

Crambessa mosaica, E. Haeckel. Das System der Medusen. Seite 622.

Cephea mosaica, Quoy et Gaimard. Voyage de l'Uranie, Zoologie; page 569, plate 85, fig. 3.

Rhizostoma mosaica, Fr. Eschscholz. System der Acalephen; page 53.

Rhizostoma mosaica, Huxley. Philosophical Transactions; pages 422, 432; plate 38; figures 26, 27; plate 39; figures 28—34.

Catostylus mosaicus, L. Agassiz. Contributions to the Natural History of the United States, IV, page 152.

Catostylus mosaicus, Grenacher und Noll. Abhandle. Senkenberg, Ges. Band X. Seite 38.

Catostylus Wilkesii, L. Agassiz. Contributions to the Natural History of United States, IV, page 152.

Crambessa mosaica, Von Lendenfeld. Ueber Wehrthiere und Nesseljellen. Zeitschrift für wissenschaftliche Zoologie. Band 38.

Crambessa mosaica, Von Lendenfeld. The Scyphomedusæ of the Southern Hemisphere. Proceedings of the Linnean Society of New South Wales ; volume ix, part 2, page 299—300.

Crambessa mosaica, Von Lendenfeld. Local Colour Variety of Scyphomedusæ. Proceedings of the Linnean Society of New South Wales ; volume ix, page 926—927.

Umbrella slightly vaulted, nearly semi-spherical, 2—3 times as broad as high, with numerous narrow marginal flaps. (To 200 ? about 24 velar flaps on each octant.) Sides of the gastro-genital cross nearly rectangular. The same breadth at the distal, and at the proximal end ; mouth-arms a little shorter than the diameter of the umbrella. Under-arms conic, pointed below, surrounded by thickly-set suctorial crisps, three times as broad as the thin and short upper-arm.

Colour.—Blue or brown, with a net of white lines on the ex-umbrella ; male genital organs—grey, green ; ovaries, deep orange-red.

Size.—Breadth of umbrella, 200—350 mm.; height of umbrella, 80—120 mm.

Ontogenesis.—Unknown.

I distinguish two varieties in this species (colour varieties of Scyphomedusæ, l.c.).

I.—CRAMBESSA MOSAICA CONSERVATIVA.

Corresponds to the descriptions of the older author, is blue and transparent.

Locality.—Port Phillip, v. Lendenfeld. Port Jackson, Quoy et, Gaimard ; Huxley ; Illawarra Lake, Drayton.

II.—CRAMBESSA MOSAICA SYMBIATICA.

The whole of the gallert is pervaded by an abundance of zooxanthella-yellow cells which renders them deep yellowish-brown and untransparent. In the young stage without yellow cells and colourless.

This variety seems to have been produced since the older observers made their collections within the last forty years.

Locality.—Port Jackson, v. Lendenfeld.

Familia LEPTOBRACHIDÆ.

Claus, 1883.

Rhizostomæ with very broad brachial disc, in which the upper-arms are more or less retracted. The slender lower-arms are ribbon-shaped elongated and furnished with three crisps ; these may occur in the distal portion only. The canal system forms a dense network nearly throughout the whole of the disc. The radial canals are numerous. The eight which lie in the primary and secondary radii are larger than the others. Sub-genital porticus continuous. Brachial disc divided from the umbrella.

Genus THYSANOSTOMA.   L. Agassiz, 1862.

The three narrow wings of the under-arm possess suctorial crisps throughout their entire length.

THYSANOSTOMA THYSANURA.   E. Haeckel.

Thysanostoma thysanura, E. Haeckel.   Das System der Medusen.   Seite 625.

Thysanostoma thysanura, R. v. Lendenfeld.   The Scyphomedusæ of the Southern Hemisphere.   Proceedings of the Linnean Society of New South Wales ; volume ix, part 2, page 303.

Umbrella nearly orbicular, two or three times as broad as high, with ninety-six marginal flaps (in each octant ten truncate rounded velar flaps between two pointed three-cornered ocular flaps). Ex-umbrella polygonal, slabbed with irregular and granulated slabs, which are scarcely half as broad as the marginal flaps. Sub-genital ostia four or six times as broad as the pillars. Arms two or three times as long as the diameter of the umbrella, very narrow, ribbon-shaped, three-winged, the same breadth nearly throughout their whole length to the rounded end.

Size.—Breadth of umbrella, 8 mm. ; height of umbrella, 3 mm.

Locality.—Australia, Museum Godeffroy.

Genus LEONURA.   Haeckel, 1879.

Suctorial crisps only at the distal ends of the lower-arms above the terminal knot. Mouth cross with 8 rays and 8 suctorial crisps in the tertiary radii, which form especial frills round the centre of the brachial disc.

LEONURA LEPTURA.   E. Haeckel.

Leonura leptura.   E. Haeckel.   Das System der Medusen.   Seite 631.

Leonura leptura, R. v. Lendenfeld.   The Scyphomedusæ of the Southern Hemisphere.   Proceedings of the Linnean Society of New South Wales ; volume ix, part 2, page 305.

Umbrella flat, orbicular, with 80 marginal flaps. In each octant 8 rectangular velar flaps, between 2 pointed triangular ocular flaps. Sub-genital ostia three times as broad as the pillars between them. Mouth-cross of the brachial disc, with frills ; the 8 rays in pairs. Arms very long and slender, ribbon-shaped, nearly three times as long as the diameter of the umbrella, a bundle of frills at the end which surrounds a pointed three cornered terminal knot.

Size.—Breadth of umbrella, 100 mm. ; height of umbrella, 50 mm.

Locality.—South Pacific Ocean, near New Zealand, Weber.

There are according to this Australian 26 species, with four varieties, which are distributed among 79 genera.

Sydney, November 30, 1885.

E. v. Lendenfeld, pinx.

EXPLANATION OF PLATE.

— o —

MEDUSÆ.

1. Phyllorhiza punctata. *R. v. L.*
2. Cyanea annaskala. *R. v. L.* Var. purpurea.
3. Monorhiza haeckeli. *Haacke.*
4. Aurelia coerulea. *R. v. L.*
5. Pseudorhiza aurosa. *R. v. L.*
6. Crambessa mosaica. *Quoy. et Gaim.* Var. conservatina.
7. Cyanea annaskala. *R. v. L.* Var. marginata.

CTENOPHORA.

8. Bolina chuni. *R. v. L.*
9. Neis cordigera. *Less.*

SPONGES.

10. Chalinopsis imitans. *R. v. L.*
11. Sycandra arborea. *Haeck.*
12. Aplysilla violacea. *R. v. L.*
13. Dendrilla rosea. *R. v. L.* Var. typica.
14. Phyllospongia foliascens. *Esp.*
15. Ianthella flabelliformis. *Anct.*
16. Rhizorchalina globosa. *R. v. L.*
17. Thallassodendron rubens. *R. v. L.* Var. digitata.
18. Sycandra ramsayi. *R. v. L.*
19. Ianthella basta. *Gray.*
20. Leucandra saccharata. *Haeck.*

The original painting is in the possession of the Linnean Society of N.S.W.

PART II.

HYDROMEDUSÆ.

A CATALOGUE

AUSTRALIAN HYDROMEDUSÆ;

By R. von LENDENFELD, Ph.D.

In the following pages a complete Catalogue of all the Australian species is given, together with all references and synonyms.

The Classification is the same which I have advocated and described in several former papers (90), (91). Type specimens of the greater number of the species described by myself are in the Australian Museum.

LITERATURE.

The Essays relating to our subject are arranged alphabetically according to the authors. The papers of the same author are arranged chronologically.

1. AGASSIZ (Alexander)—1865.
   North American Acalephæ. Illustrated Catalogue of Comparative Zoology at Harvard College, Cambridge, Mass., No. II.

2. AGASSIZ (Louis)—1860–1862.
   Contributions to the Natural History of the United States of America.

3. ALDER (T.)—1857.
   A Catalogue of the Zoophytes of Northumberland and Durham. Transactions of the Tyne Naturalists' Field Club for 1857.

4. ALLMAN (George James)—1871–1872.
   A Monograph of the Gymnoblastic or Tubularian Hydroids; Vols. I, II, Roy. Society for 1870–1871.

5. ALLMAN (George James)—1876.

New Genera and Species of Hydroida. Journal of the Linnean Society of London; Zoology, Vol. XII.

6. ALLMAN (George James)—1877.

Report on the Hydroida collected during the Exploration of the Gulf Stream; by L. F. de Pourtale's Assistant, United States Coast Survey. Cambridge, Mass., 1877.

7. ALLMAN (George James)—1883.

Report on the Hydroida. First part; the Plumularidæ. Report on the Scientific Results of the Voyage of H.M.S. "Challenger"; Zoology, Vol. VII.

8. BALE (William M.)—1884.

On the Hydroida of South-eastern Australia. Journal of the Microscopical Society of Victoria; Vol. II, Part I.

9. BALE (William M.)—1884.

Catalogue of the Australian Hydroid Zoophytes. Sydney, 1884.

10. BENEDEN (P. J. von)—1844.

Mémoire sur les Campanulaires de la côte d'Ostende considérés sous le rapport physiologique, embryogénique et zoologique. Nouveaux Mémoires de l'Académie de Bruxelles; Tome 17.

11. BENEDEN (P. J. von)—1847.

Sur la réproduction des Campanulaires. Bulletin de l'Académie de Belgique. Tome pour 1847.

12. BENEDEN (P. J. von)—1866.

Recherches sur la Faune littorale de Belgique; Polypes; Bruxelles 1886.

13. BERGH (R. S.)—1878.

Nogle Bidrag til de athecate Hydroiders Histologi Videns Rabelige Meddelser fro naturhistoriske; Torening i Kjöbenhaven, 1878.

14. BERGH (R. S.)—1879.

Studien ueber die erste Entwickelung des Eies von Gonothyrea Loveni. Morphologisches Tahrbuch. Band V.

15. BLAINVILLE (Henri Marie Ducrotay de)—1834-1837.

Manuel d'Actinologie ou de Zoophytologie. Paris, 1834-1837.

16. BOHM (R.)—1878.

Helgolander Leptomedusen. Jenaische Zeitschrift für naturwissenschaft. Band XII.

17. BUSK (George)—1851.

On Sertularian Zoophytes of South Africa. Report of the British Association for 1851.

18. Busk (George)—1852.

An Account of the Polizoa and Sertularian Zoophytes collected in the voyage of the "Rattlesnake" on the Coast of Australia and the Louisiade Archipelago. Narrative of the Voyage of H.M.S. "Rattlesnake": App. iv.

19. Carter (H. T.)—1873.

On New Species of Hydractinidæ. Annals and Magazine of Natural History; 4th series; Vol. XI.

20. Ciamician (J.)—1878.

Zur Frage ueber die Enstehung der Geschlechtsstoffe bei den Hydroiden. Zeitschrift für wissenschaftliche Zoologie; Band XXX, seite 501.

21. Ciamician (J.)—1879.

Ueber den feinern Bau und die Entwickelung von Tubularia mesembryanthemum, Allman. Zeitschrift für wissenschaftliche Zoologie; Band XXXII, seite 323.

22. Ciamician (J.)—1879.

Ueber Lafoëa parisitica, nov. spec. Zeitschrift für wissenschaftliche Zoologie; Band XXXIII, seite 673.

23. Claus (Carl)—1880.

Grundzüge der Zoologie. IV Auflage, seite 218.

24. Claus (Carl)—1881.

Beiträge zur Kenntuiss der Gerijonopsiden und Encopiden Entwickelung. Arbeiten aus dem Zoologischen Institute der Universität Wien.

25. Coughtrey.—1876–1877.

New Zealand Hydroida. Transactions of the New Zealand Institute. Vols. VII, VIII.

26. Coughtrey.—1876.

Critical Notes on the New Zealand Hydroida. Annals and Magazine of Natural History. 4th series; Vol. XVII.

27. Dana (James)—1846.

Exploring Expedition of the United States. Philadelphia, Mass., 1846.

28. Dujardin (Feé)—1843.

Observations sur un nouveau genre de Medusaires (Cladonema) provenant de la Métamorphose des Sycorynes. Annales des Sciences Naturelles; 2ᵉ série, 1843.

29. Dujardin (Feé)—1845.

Mémoire sur le développement des Medusaires et des Polypes Hydraires. Annales des Science Naturelles. 2ᵉ série, 1845.

30. Du Plessis (G.)—1880.

Catalogue provisoire des Hydroides Medusipares observés durant l'hiver, 1879-80, à la station Zoologique de Naples. Mittheilungen der Zoologischen Station in Neapel. Band II, seite 143.

31. EDWARDS (Milne. Blanchard et Quatrefage)—1849.
Le Règne Animal de Cuvier, illustré. Zoophytes. Paris, 1849.

32. EDWARDS (Milne) et HAIME—1848-1852.
Recherches sur le Polypiers. Annales des Sciences Naturelles. 3² série. Tome XIII.

33. EDWARDS (Milne) et HAIME—1860.
Histoire Naturelle de Corallaires. Tome III.

34. EIMER (Theodor)—1879.
Die Medusen Physiologisch und Morphologisch auf ihr Nervensystem untersucht. Tübingen, 1878 (recte 1879).

35. ELLIS (John)—1754.
Essay towards a Natural History of the Corallines found on the Coast of Great Britain and Ireland. London, 1755.

36. ELLIS (J.) and SOLANDER—1786.
The Natural History of Many Curious and Uncommon Zoophytes. London, 1786.

37. ESCHSCHOLTZ (Fr.)—1829.
System der Acalephen. Berlin, 1829.

38. ESPER (E. J. C.)—1805-1830.
Die Pflanzenthiere. Nuernberg, 1805-1830.

39. EYDOUX (X.) et L. SOULEYET—1841-1852.
Zoologie. Voyage autour du monde en 1836-1837 sur la corvette "La Bonite." Paris, 1841 1852. Tome II.

40. FLEMING (John)—1822.
Natural History of British Animals. London, 1822.

41. FORBES (Edward)—1848.
A Monograph of the British naked-eyed Medusæ. Roy. Society for 1848.

42. FRAIPONT (Julien) 1879.
Recherches sur l'Organisation histologique et le Développement de la Campanularia angulata. Archives de Zoologie expérimentale et générale. Tome VIII.

43. GEGENBAUR (Carl)—1854.
Zur Lehre von Generations, wechsel und der Fortpflanzung bei Medusen und Polypen. Nuernberg, 1854.

44. GEGENBAUR (Carl)—1878.
Grundriss der vergleichenden Anatomie. Leipzig, 1878.

45. GMELIN—1789.
Linné. Systema naturæ, XIII ; Auflage.

46. GRAY (J. E.)—1843.

Materials towards a Fauna of New Zealand : additional radiate animals. Dieffenbach. Travels in New Zealand. Vol. II.

47. GRAY (J. E.)—1868.

The Ceratellidæ. Proceedings of the Zoological Society of London. Vol. VIII.

48. GROBBEN (Carl)—1875.

Ueber Podocoryne carnea, Sars. Sitzungsberichte der Kaiserlichen Akademie der Wissenschaften. Wien. Band 72.

49. HAACKE (Wilhelm)—1880.

Zur Blastologie der Gattung Hydra, specielle und generelle Studien. Tenaische Zeitschrift für Naturwissenschaft. Band XIV, seite 133.

50. HAECKEL (Ernst H.)—1879.

Das System der Medusen. Erster Theil einer Monografie der Medusen. Jena, 1879.

51. HAECKEL (Ernst H.)—1881.

Die Tiefseemedusen der "Challenger"-Reise. Erste Hälfte des zweiten Theiles einer Monografie der Medusen. Der Organismus der Medusen. Zweite Hälfte des zweiten. Theiles einer Monografie der Medusen. Jena, 1881. Report of the Voyage of H.M.S. "Challenger"; Zoology, Vol. IV.

52. HALLEY (J. J.)—1879.

On Tubularia Ralphii, manuscript. Read before the Microscopical Society of Victoria. June, 1879.

53. HAMANN (Otto)—1882.

Studien ueber coelenteraten Jenaische Zeitschrift für Naturwissenschaft. Band XV, seite 545.

54. HAMANN (Otto)—1882.

Der Organismus der Hydroidpolypen. Jenaische Zeitschrift für Naturwissenschaft. Band XV, seite 473.

55. HAMANN (Otto)—1882.

Die Entstehung und Entwickelung der grünen Zellen bei Hydra. Zeitschrift für wissenschaftlicke Zoologie. Band XXXVII, seite 457.

56. HELLER (Cam.)—1868.

Die Zoophyten und Echinodermen des Adriatischen Meeres. Wien, 1868.

57. HERTWIG (Oscar und Richard) 1878.

Das Nervensystem und die Sinnesorgane der Medusen. Jena, 1878.

58. HERTWIG (Oscar und Richard)—1878.

Der Organismus der Medusen. Jena, 1878.

59. HERTWIG (Oscar and Richard)—1880.
Die Actinien anatomisch und histologisch mit besonderer Berücksichtigung des Nervenmuskelsystems untersucht. II. Theil. Jenaische Zeitschrift für Naturwissenschaft. Band XIV, seite 39.

60. HINCKS (Thomas)—1861.
On new Australian Hydrozoa. Annals and Magazine of Natural History. 3rd series, Vol. VII.

61. HINCKS (Thomas)—1862.
A Catalogue of the Zoophytes of South Devon and South Cornwall. Annals and Magazine of Natural History. 3rd series, Vol. IV.

62. HINCKS (Thomas)—1863.
On new British Hydroida. Annals and Magazine of Natural History. 3rd series, Vol. XI.

63. HINCKS (Thomas)—1868.
A History of the British Hydroid Zoophytes. 2 Vols. London, 1868.

64. HUTTON (Francis)—1872.
New Zealand Sertularians. Transactions of the New Zealand Institute. Vol. V.

65. JICKELI (Carl F.)—1883.
Der Bau der Hydropolypen, I, II. Morphologisches Jahrbuck. Band VIII, seites 373, 580.

66. JICKELI (Carl F.)—1882.
Ueber Hydra. Zoologischer Anzeiger. Band V, seite 91.

67. JOHNSTONE (George)—1849.
A History of British Zoophytes. 2nd edit. London, 1847.

68. KEFERSTEIN (und E. Ehlers)—1861.
Zoologische Beiträge, gesammelt im Winter, 1859-60, in Neapel und Messina. Leipzig, 1861.

69. KENT (W. Saville)—1871.
On Corals. Proceedings of the Geological Society of London for 1871.

70. KERCHNER (Ludwig)—1880.
Zur Entwickelungsgeschichte der Hydra. Zoologischer Anzeiger. Band III, seite 454.

71. KIRCHENPAUER (G. H.)—1864.
Neue Sertularien aus verschiedenen Hamburgischen Sammlungen nebst allgemeinen Bemerkungen über Lamaroux's Gattung Dynamena. Verhandlungen der Kaiserlichen, Leopoldins—carolinischen deutschen. Akademie der Naturforcher. Band xxxi.

72. KIRCHENPAUER (G. H.)—1872.
Ueber die Hydroidenfamilie Plumulariadæ; einzelne Gruppen derselben, und ihre Fruchtbehälter. Abhandlungen aus dem Gebiete der Naturwissenschaften, herausgegeben von dem naturwissenschaftlichen Verein in Hamburg. Band v, Abtheilung 3.

73. KIRCHENPAUER (G. H.)—1876.

Ueber die Hydroidenfamilie Plumularidæ, einzelne Gruppen derselben, und ihre Fruchtbehälter. Abhandlungen aus dem Gebiete der Naturwissenschaften, herausgegeben von Naturwissenschaftlichen Verein zu Hamburg—Altona. Band vi, Abtheilung 2.

74. KLAATSCH (H.)—1884.

Beitrage zur genaueren Kenntniss der Campanularien. Morphologisches Jahrbuch. Band ix, seite 534.

75. KLEINENBERG (Nikolaus)—1872.

Hydra, eine anatomischentwickelungsgeschichtliche Untersuchung. Leipzig, 1872.

76. KLEINENBERG (Nikolaus)—1881.

Ueber die Entstehung der Eier bei Eudendrium. Zeitschrift für wissenschaftliche Zoologie. Band xxxv, seite 326.

77. KÖLLIKER (Albert von, C. Gegenbauer und F. Müller).

Berichte ueber einige im Herbste 1852 in Messina angestellte, verglcichendanatomische Untersuchungen. Zeitschrift für wissenschaftliche Zoologie. Band iv.

78. KOREN (und Daniellsen)—1875.

Fauna littoralis Norvægiæ. Theil iii.

79. KOROTNEFF (A.)—1880.

Versuch einer vergleichenden Theorie der cœlenteraten (Myriothella und Hydra russich). Moskau, 1880.

80. KOROTNEFF (A.)—1882.

Zur Kenntniss der Embryologie von Hydra. Zeitschrift für wissenschaftliche Zoologie. Band xxxviii, seite 314.

81. LAMARCK (Jean de)—1817.

Histoire Naturelle des animaux sans vertèbres. Tome ii. Paris, 1817.

82. LAMARCK (Jean de)—1835.

Histoire Naturelle des animaux sans vertèbres. Deuxième édition, par MM. G. P. Deshayes et H. Milne-Edwards. Tome ii, Histoire des Polypes. Paris, 1835.

83. LAMOUROUX (T. V. F.)—1816.

Histoire des Polypiers Coralligènes flexibles vulgairement nommés. Zoophytes. Caen, 1816.

84. LAMOUROUX (T. V. F.)—1821.

Exposition Methodique des generes de l'ordre des Polypiers. Paris, 1821.

85. LAMOUROUX (T. V. F.)—1824.

Bory de Saint Vincent et End. Deslongchamps Histoire Naturelle des Zoophytes ou animaux rayonnés. Tome II. Encyclopédie Methodique.

86. LENDENFELD (R. von)—1883.

Ueber eine eigenthümliche Art des Sprossen bildung bei Campanulariden. Zoologischer Anzeiger. Band VI, seite 71.

87. LENDENFELD (R. von)—1883.

Ueber das Nervensystem der Hydroidpolypen. Zoologischer Anzeiger. Band VI ; Seite 69.

88. LENDENFELD (R. von.)—1883.

Ueber Wehrthiere und nesselzellen. Zeitschrift für wissenschaftliche Zoologie. Band XXVIII, seite 335. Annals and Magazine of Natural History. Vol. XII, page 250.

89. LENDENFELD (R. von)—1883.

Encopella Campanularia. Zeitschrift für wissenschaftliche Zoologie. Band XXVIII, seite 497.

90. LENDENFELD (R. von)—1884.

Das System der Hydromedusen. Zoologischer Anzeiger. Band VII, seite 173.

91. LENDENFELD (R. von)—1884.

The Australian Hydromedusae, Part I. The Classification of the Hydromedusae. Proceedings of the Linnean Society of New South Wales. Vol. IX, page 206.

92. LENDENFELD (R. von)—1884.

The Australian Hydromedusae, Part II. The first Sub-order; Hydropolypinae; the Families Hydridae, Claridae, Myriothelidae, and Eudendridae. Proceedings of the Linnean Society of New South Wales. Vol. IX, page 345.

93. LENDENFELD (R. von)—1884.

The Australian Hydromedusae, Part III. The first Sub-order Hydropolypinae ; the Blastopolypidae. Proceedings of the Linnean Society of New South Wales. Vol. IX, page 401.

94. LENDENFELD (R. von)—1884.

The Australian Hydromedusae, Part IV. The Graptolithidae, Plumularidae, and Dicorymidae. Proceedings of the Linnean Society of New South Wales. Vol. IX, page 467.

95. LENDENFELD (R. von) -1884.

The Australian Hydromedusae, Part V. The Hydromedusinae, Hydrocorallinae, and Trachimedusae. Proceedings of the Linnean Society of New South Wales. Vol. IX, page 581.

96. LENDENFELD (R. von)—1884.

Sarsia radiata und der Flexor ihres Polypen. Zoologischer Anzeiger. Band VII, seite.

97. LENDENFELD (R. von)—1884.

On Muscular tissue in Hydroid Polyps. Proceedings of the Linnean Society of New South Wales. Vol. IX, page 635.

98. LENDENFELD (R. von)—1884.

Addenda to the Australian Hydromedusæ. Proceedings of the Linnean Society of New South Wales. Vol. IX, page 908.

99. LESSON (R. P.)—1829.

Zoologic. Voyage de la " Coquille." Paris, 1829.

100. LESSON (R. P.)—1843.

Prodrôme des Acalèphes. Paris, 1843.

101. LEUCKART (Rudolph)—1851.

Ueber den Polymorphismus der Individuen oder die Erscheinungen der Arbeits theilung in der natur. Giessen, 1851.

102. LESUEUR (C. A.)—1809.

Recueil des Planches (inédites) des Méduses.

103. LISTER—1834.

On the Hydroida. Philosophical Transactions of the Royal Society of London for 1834.

104. McCoy (Frederic)—1874.

Prodromus of the Palæontology of Victoria ; or, Figures and Descriptions of Victorian Organic Remains. Geological Survey of Victoria. Decade I.

105. McCoy (Frederic)—1875.

Prodromus of the Palæontology of Victoria ; or, Figures and Descriptions of Victorian Organic Remains. Geological Survey of Victoria. Decade II.

106. McCoy (Frederic)—1878.

Prodromus of the Palæontology of Victoria ; or, Figures and Descrip tions of Victorian Organic Remains. Geological Survey of Victoria. Decade V.

107. MARSHALL (William)—1882.

Ueber einige Lebenserscheinungen der Süsswasserpolypen, und ueber eine neue Form von Hydra viridis. Zeitschrift für wissenschaft- liche Zoologie. Band XXXVII, seite 664.

108. METSCHNIKOFF (Elias)—1883.

Untersuchungen über die introcelluläre Verdauung bei wirbellosen Thieren. Arbeiten aus dem zoologischen Institut der Universitat Wien. Band V, seite 171.

109. MEREJKOVSKY (C. von)—1882.

Structure et développement des nématophores chez les Hydroides. Archives de Zoologie expérimentale et générale. Tome X, page 583.

110. MOSELEY (H. N.)—1877.

On the Structure of a species of Millepora, occurring at Tahiti. Trans- actions of the Royal Society of London for 1877, page 117.

111. Moseley (H. N.)—1878.

On the Structure of the Stylasteridæ, a group of the Hydroid stony corals. Transactions of the Royal Society of London for 1878, page 425.

112. Oken (Lorenz)—1815.

Zoologie, Lehrbuch der Naturgeschicte. Band III.

113. Pallas (P. S.)—1766.

Elenchus Zoophytorum. Hogae, 1766.

114. Parker (J. Jeffrey)—1880.

On the Histology of Hydra fusca. Transactions of the Royal Society of London for 1880, page 61, No. 200.

115. Peron (et Lesueur)—1809.

Tableau des Méduses. Paris, 1809, Annales du Museum d'histoire Naturelle No. XIV.

116. Poeppig—1876.

Manuscript (73).

117. Quatrefages (A. de)—1843.

Mémoire sur la Syhydre parasite (Hydractinia echinata). Annales des Sciences Naturelles. 2ᵉ série, Vol. XX.

118. Quoy (T. R. C. et P. Gaimard)—1824.

Zoologie. Voyage autour du monde sur les corvettes "l'Uranie" et "la Physicienne."

119. Sars (Martin)—1850.

Reise i Lofoten og Finmarken. Nyt Magazine für Naturvidskab, 1850.

120. Sars (Martin)—1857.

Bidrag til Kundkabn om Middlehavets Littoral Fauna, 1857.

121. Sars (Martin)—1862.

Bemaerkinger ouer fire norske Hydroider. Videnskab Forhandlungen für 1862.

122. Sars (G. O.)—1873.

Bidrag til Kundkabn om Norges Hydroida, 1873.

123. Saunders—Manuscript in litteris.

124. Schulze (F. E.)—1871.

Cordylophora lacustris. Leipzig, 1871.

125. Schulze (F. E.)—1873.

Syncoryne Sarsii Lovén und die Zugehörige Meduse Sarsia tubulosa Lesson. Leipzig, 1873.

126. Schweiger (A. F.)—1819.

Beobachtungen auf Naturhistorischen Reisen. Berlin, 1819.

127. THOMPSON (D'Arcy, W.)—1879.

New and rare Hydroid Zoophytes (Sertularidæ and Thuiariidæ) from Australia and New Zealand. Annals and Magazine of Natural History. 5th series, Vol. III, page 97.

128. TOURNEFORT (Jas. Pit).—1700.

Institutiones rei herbariæ. Paris, 1700.

129. TREMBELEY—1774.

Memoire pour servir à l'histoire d'un genre de Polypes d'eau douce à bras en forme de cornes. Paris, 1744.

130. VARENNE (André de)—1882.

Recherches sur les Polypes Hydraires (Reproduction et développement). Archives de Zoologie experimentale et générale. Tome X.

131. WEISMANN (August)—1883.

Die Entstehung der Sexualzellen bei den Hydromedusen. Jena, 1883.

132. WOODS (Tenison)—1878.

On a new genus of Milleporidæ. Proceedings of the Linnean Society of New South Wales. Vol. III.

133. WOODS (Tenison)—1879.

On the Anatomy of Distichopora. Transaction of the Royal Society of New South Wales. Vol. XIII.

134. WRIGHT (Strethill)—1858.

Observations on British Zoophytes. Edinburgh new Philosophical Journal. Vol. VII.

135. ZITTEL (Carl)—1876-1880.

Handbuch des Paleontologie. I. Band. Paleozoologie.

ORDO HYDROMEDUSÆ.

Carus, 1863.

POLYPOMEDUS.E, without gastral filaments. The polypoid forms rarely solitary; they mostly form colonies, and then always possess a chitinous or calcareous skeleton. The trophosomes are destitute of an œsaphagal tube. Some of the tubes are often converted into mouth and tentacleless sexual or defensive zooids. The medusoid forms are cycloneur. (2) (23) (13) (44) (59) (90) (91) (101) (135).

I. SUB-ORDO.—HYDROPOLYPINÆ.

V. Lendenfeld, 1884.

Polyps, or polyp colonies, the sexual products of which are matured in the ordinary alimentary polyp or in zooids which are converted into mouthless generative zooids, polypostyles. (4) (20) (29) (54) (65) (87) (130) (131).

I.—Familia HYDRIDÆ.

Huxley, 1856.

Solitary polyps; the sexual products are matured in the gastral wall. Without hydrothecæ.

1. Genus HYDRA. Linné.

Nonsexual propagation by budding. (49) (53) (55) (65), (66) I 391. (70) (75) (79) (80) (107) (114).

1. **Hydra obigactis.** Pallas. (9) 187, (62) 315, (67) 124, (92) 318.
   Hydra fusca (45) 1320.
   Long-armed fresh-water polyps (38) 16.
South-east Australia (Victoria).

2. **Hydra viridis.** V. Linné. (26) 24, (15) 1320, (67) 121, (92) 318.
   Hydra viridissima (113) 31.
   Polype verda (127) 22.
New Zealand.

II.—Familia CLAVIDÆ.

V. Lendenfeld, 1884.

Forming colonies ; polyps alike with scattered tentacles, and all maturing sexual cells on hollow tentacular processes. Without a hydrotheca.

I.—Sub-familia CLAVINÆ. V. Lendenfeld, 1884.

With scattered filiform tentacles.

2. Genus CLAVA. Gmelin.

Club-shaped clavinæ, without distinct limit between polyp and hydrocaulus. (131) 21.

3. **Clava simplex.** V. Lendenfeld. (92) 349.
East coast of Australia (Port Jackson).

II.—Sub-familia CORYNINÆ. V. Lendenfeld, 1884.

With scattered capitate tentacles. (65) II 610, (131) 49.

III.—Familia MYRIOTHELIDÆ.

Allman, 1872.

Solitary with scattered capitate tentacles. The sexual products are matured on branched tentacular processes of the body-wall. Without hydrothecæ. (79).

IV.—Familia EUDENDRIDÆ.

Allman, 1872.

Forming colonies ; polyps without hydrothecæ with one verticil of filiform tentacles. The sexual products are matured on tentacular processes of the gastral wall ; during their growth the polyps loose their tentacles and become polypostyle. The phylogenesis of the polypostyles is continually being repeated.

3. Genus EUDENDRIUM. Ehrenberg.

The polyps terminal on the branches of the ramified colony. A ring of gland cells at the base of the polyps. (54) 522, (65) I 376, (76) (92) 350, (131) 91.

4. **Eudendrium pusillum.** V. Lendenfeld. (92) 352.
East coast of Australia (Port Jackson).

5. **Eudendrium generalis.** V. Lendenfeld. (92) 351.
South coast of Australia (Port Phillip).

V. Familia BLASTOPOLYPIDÆ.

V. Lendenfeld, 1884.

Forming colonies, with differentiated alimentary and mouthless generative zooids. The latter are transformed polyps, polypostyles, and only in them the sexual products are matured.

I.—Sub-familia CORDYLOPHORINÆ. V. Lendenfeld, 1884.

Alimentary zooids of the colony with scattered filiform tentacles, without hydrothecæ. (65) II 601, (124) (131) 29.

II.—Sub-familia BIMERINÆ. V. Lendenfeld. 1884.

Alimentary zooids, with one verticil of filiform tentacles, without hydrothecæ.

III.—Sub-familia CAMPANULARINÆ. V. Lendenfeld, 1884.*

Alimentary zooids, with one verticil of filiform tentacles enclosed in a radially symmetrical hydrotheca, terminal on the branches.

4. Genus MONOSKLERA. V. Lendenfeld.

The stems consist of wedge-shaped internodes. The perisarc much thicker on one side than the other; the short hydrocauli originate on the terminal end of the internodes.

6. **Monosklera pusilla.** V. Lendenfeld. (98)

South coast of Australia (Port Phillip).

5. Genus LAOMEDEA. Lamouroux.

The hydrocauli appear as long branches of a stem, and never arise from the hydrorhiza; the internodes are cylindrical, the hydrothecæ not, operculate.

7. **Laomedea antipathies.** Lamouroux. (15) 474, (83) 206, (85) 481, (93) 403.

Campanularia antipathes (9) 52.

Sertularia antipathes (82) 138.

Australia.

8. **Loamedea Torresii.** Busk. (17) 402, (93) 403.

Campanularia Torresii (9) 52.

North coast of Australia (Torres Straits).

9. **Laomedea reptans.** Lamouroux. (84) 14, (85) 483, (93) 403.

Campanularia reptans (9) 53, (15) 473.

Sertularia reptans (82) 139.

North-west coast of Australia (Lewin's Land).

10. **Laomedea Lairii.** Lamouroux. (83) 207, (84) 14, (85) 482, (93) 403.

Campanularia Sarsii (9) 53, (82) 153.

South coast of Australia (Port Phillip, Portland).

---

*It is probable that this whole sub-familia must be united with the campanulinidæ and placed in the sub-ordo Hydromedusinæ.

11. **Laomedea marginata.** V. Lendenfeld. (93) 404
    Campanularia marginata (9) 54.
South coast of Australia (Portland).

12. **Laomedea rufa.** V. Lendenfeld. (93) 404.
    Campanularia rufa (9) 54.
East coast of Australia (Holborn Island).

13. **Laomedea undulata.** V. Lendenfeld. (93) 404.
    Campanularia undulata (9) 55, (81) 135.
    Clythia undulata (84) 202, (117) 194.
East coast of Australia (Port Jackson).

6. Genus LAFOËA. Lamouroux.

Polyps distributed along the stem and branches, with short peduncles, with a thin-walled cylindrical hydrotheca. (20), (65) II 629.

14. **Lafoëa cylindrica.** V. Lendenfeld. (98).
East coast of New Zealand (Bay of Islands).

15. **Lafoëa fruticosa.** Sars. (9) 64, (68), (93) 404.
    Callicella fruticosa (61) 293.
    Campanularia gracillima (3) 129.
South coast of Australia (Bass' Straits).

4. Sub-familia Sertularinæ. V. Lendenfeld, 1884.

Alimentary zooids, invested by bilateral symmetrical hydrothecæ, which are more or less adnate to the stem, and never possess separate hydrocauli. (71).

7. Genus LINEOLARIA. Hincks.

Polyps sessil, stem creeping. No operculum.

16. **Lineolaria spinulosa.** Hincks. (9) 61, (60), (93) 405.
South coast of Australia (Port Phillip, Portland).

17. **Lineolaria flexuosa.** Bale. (9) 62, (93) 405.
South coast of Australia (Port Phillip).

8. Genus SYNTHECIUM. Allman.

The gonophores situated in the interior of ordinary hydrothecæ. Hydrothecæ adnate, operculate.

18. **Synthecium elegans.** Allman. (4) II, 229, (98).
    Australia.

9. Genus SERTULARELLA. Gray.

Hydrothecæ adnate with a composite operculum in two opposite series, disposed regularly opposite or alternate. (130) 169.

19. **Sertularella microgona.** V. Lendenfeld. (93) 416.
South coast of Australia (Port Phillip).

20 **Sertularella polyzonias.** Gray. (9) 104, (63) 235, (93) 417.
Cotulina polyzonias (2) 356.
Great tooth Coralline (35) 5.
Sertularia Ellisii (82) 142.
Sertularia ericoides (113) 127.
Sertularia flexuosa (45).
Sertularia Hibernica (67) 128.
Sertularia pinnata. Templeton.
Sertularia polyzonias (38) T. VI. figs 1-6, (67) 61, (82) 112, (83) 190
Sertularia simplex (25), (64).
South coast of Australia (Port Phillip).
New Zealand.

21. **Sertularella indivisa.** Bale. (8) 24, (9) 105, (93) 417.
South Coast of Australia : (Victoria and South Australia.)

22. **Sertularella solidula.** Bale. (8) 24, (9) 106, (93) 417.
South coast of Australia (Port Phillip).

23. **Sertularella macrotheca.** Bale. (8) 25, (9) 107, (93) 417.
South coast of Australia (Western Port in Victoria).

24. **Sertularella lævis.** Bale. (8) 24, (9) 107, (93) 417.
South coast of Australia (Port Phillip).

25. **Sertularella pygmæa.** Bale. (8) 25, (9) 108, (93) 417.
South coast of Australia (Victoria and South Australia) New Zealand.

26. **Sertularella Johnstoni.** Allman. (5) (9) 109, (93), 418.
Sertularia Johnstoni (25), (46), (64).
South coast of Australia (Victoria and South Australia), Tasmania, New Zealand.

27. **Sertularella divaricata.** Busk. (9) 110. (18), 388, (93) 418.
East coast of Australia (Port Stephens). South coast of Australia (Bass' Straits).

28. **Sertularella neglecta.** Thompson. (9) 110, (93) 418, (127) 100.
South coast of Australia (Victoria, South Australia).

29. **Sertularella ramosa.** Thompson.* (9) 111, (93) 418, (127) 102.
South coast of Australia (Bass' Straits).

10. Genus DIPHASIA. Agassiz.

The trophosomes possess an internal operculum. Polypostyles different in the two sexes. The female gonophores with marsupium.

* Thompson is not sure whether this species described by him comes from Bass' Straits (127) 102.

B

30. **Diphasia pinnata.** Agassiz.   (2) 355, (9) 98, (63), 255, (93) 415.
    Diphasia nigra (2) 355.
    Nigellastrum nigrum (112) 93.
    Nigellastrum pinnatum (112) 93.
    Sertularia fuscescens (45) IV 677, (83) 195, (85) 683.
    Sertularia nigra (67) 68, (113) 135.
    Sertularia pinnata (67) 69, (113) 136.
    East coast of Australia (Port Jackson).

31. **Diphasia attenuata.** Hincks.   (9) 100, (63) 247, (93) 415.
    Sertularia attenuata (62) 298.
    Sertularia pinaster (67) 72.
    Sertularia rosacea (35) 9, (67) 470.
    South coast of Australia (Port Adelaide).

32. **Diphasia digitalis.** Bale.   (9) 101, (93) 415.
    Sertularia digitalis (18) 393.
    North coast of Australia (Torres Straits).

33. **Diphasia mutulata.** Bale.   (9) 101, (93) 416.
    Sertularia mutulata (18) 391.
    North coast of Australia (Torres Straits).

34. **Diphasia subcarinata.** Bale.   (9) 102, (93) 416.
    Sertularia subcarinata (18) 390.
    South coast of Australia (Victoria, Bass' Straits), east coast of Australia (Port Stephens).

35. **Diphasia rectangularis.** V. Lendenfeld.   (98).
    North coast of Australia (Torres Straits).

36. **Diphasia symmetrica.** V. Lendenfeld (93) 414. .
    East coast of New Zealand (Timaru).

11. Genus PASYTHEA. Lamouroux.

The trophosomes arranged in sets at some distance apart.

37. **Pasythea quadridentata.** Lamouroux.   (9) 112, (83) 156, (84) 9,
    (85) 603, (93) 419
    Sertularia quadridentata (36) 57, (38)
    Suppl. II, tab. 32, figs. 1–5 (45) 3883, (82) 150.
    Tuliparia quadritentata (15) 485.
    East coast of Australia (Port Stephens, Fitzroy Islands).

38. **Pasythea hexodon.** Busk.   (9) 113, (18) 395, (93) 419.
    North coast of Australia (Cumberland Island).

12. Genus IDIA. Lamouroux.

Colonies pinnately branched. The trophosomes form two continuous series in contact with each other along the front of the stem.

39. **Idia pristis.** Lamouroux. (9) 113, (83) 200, (84) pl. V., fig. 5 (85) 462, (93) 419.

Sertularia pristis. (18) 390.

North coast of Australia (Torres Straits), east coast of Australia (Fitzroy Islands, Albany Passage).

13. Genus THUIARIA. Fleming.*

Trophosomes biserial, not in pairs, usually more or less immersed. The series differently dense.

40. **Thuiaria fenestrata.** Bale. (9) 116, (93) 420.

Salacia tetrocytharia (83) 214, (84), g. pl. 15, (85) 673.
Sertularia crisioides (18) 389.
North coast of Australia (Torres Straits), east coast of Australia (Albany Passage).

41. **Thuiaria quadridens.** Bale. (9) 119, (93) 420, (98).

East coast of Australia (Port Curtis, Holborn Island), New Zealand.

42. **Thuiaria lata.** Bale. (8) 26, (9) 120, (93) 420.

East coast of Australia (Port Stephens), south coast of Australia (Port Phillip, Western Port).

VI.—Familia RHABDOPHORA.
Allman, 1872.

Possessing a chitinous endo and exo-skeleton ; the former rod-shaped ; colonies free swimming ; probably extinct.

I. Group.—GRAPTOLOIDEA.
Lapworth.

Hydrosome developed from a sicula, every canal containing cœnosarc, bears only one row of cells. Axis (virgula) on the dorsal side in a furrow of the inner lamina.

A.—MONOPRIONIDÆ.

Hydrothecæ in one row opposite the axis.

I.—Sub-familia MONOGRAPTINÆ. V. Lendenfeld.
= MONOGRAPTIDÆ. Lapworth.

Developed one-sided ; pointed ends of the sicula pointing upwards, united with the dorsal margin of the proximal end of a single or composite hydrosome.

---

* I had no opportunity to examine sufficiently well preserved gonophores of Thuiaria. It appears highly probable that this genus belongs to the Hydromedusinæ.

II.—Sub-familia LEPTOGRAPTINÆ. V. Lendenfeld.
LEPTOGRAPTIDÆ. Lapworth.

Hydrosome bilateral, with irregular branches. Cells apart, just touching. Sicula persistent in the axilar. The broad part forming the proximal end of the hydrosome.

III.—Sub-familia DICHOGRAPTINÆ. V. Lendenfeld.
DICHOGRAPTIDÆ. Lapworth.

Bilateral. Branches of regular cells, very dense, rectangular. Sicula persistent, its point at the proximal end of the hydrosome.

14. Genus Didymograpsus. McCoy.

Only two simple branches, without funiculus. Sicula axilar, with the point turned upwards.

43. **Didymograpsus fruticosus.** Hall. (94) 468, (104) 13.
Bendigo, a.o.p. Victoria.

44. **Didymograpsus quadribrachiatus.** Hall. (94) 468, (104) 15.
Victoria.

45. **Didymograpsus Bryonoides.** Hall. (94) 469, (104) 16.
Victoria.

46. **Didymograpsus octabrachiatus.** Hall. (94) 469, (104) 17.
Victoria.

47. **Didymograpsus logani.** Hall. Var. Australis. McCoy. (94) 469, (104) 18.
Castlemaine, Kangaroo Creek, a.o.pl. Victoria.

48. **Didymograpsus extensus.** Hall. (94) 469, (105) 29.
Bendigo, Victoria.

49. **Didymograpsus caducens.** Salter. (94) 469, (105) 30.
Castlemaine, a.o.p. Victoria.

50. **Didymograpsus gracilis.** Hall. (94) 469, (105) 35.
Bulla, Victoria.

51. **Didymograpsus thureani.** McCoy. (94) 469, (105) 39.
Sandhurst, Victoria.

52. **Didymograpsus headi.** McCoy. (94) 469 (105) 40.
Victoria.

15. Genus CLADOGRAPSUS. McCoy.

Stem simple below, with two rows of cells and mid-rib as in Diplograpsus; dividing above into branches with one row of cells only; cells excavated in the margin as in Climacograptus; without distinct tubes.

53. **Cladograpsus ramosus.** Hall. (94) 469, (105) 33.
Bulla, Victoria.

54. **Cladograpsus furcatus.** Hall. (94) 470, (105) 37.
Bendigo, Victoria.

IV.—Sub-familia DICRANOGRAPTINÆ. Von Lendenfeld.
DICRANOGRAPTIDÆ. Lapworth.

Hydrosome consists of two originally dorsally united axes. Cells overlapping. Exterior part indented. Broad end of the sicula on the proximal end of the hydrosome.

B. DIPRIONIDÆ.

Cells in two rows ; axis central.

V.—Sub-familia DIPLOGRAPTINÆ. Von Lendenfeld.
DIPLOGRAPTIDÆ. Lapworth.

Hydrosome consists of two branches dorsally joined. Sicula imbedded. The broad part forming the proximal end of the hydrosome.

16. Genus DIPLOGRAPSUS. McCoy.

Stems simple, straight, with a slender central axis ; cells oblique in two rows, alternating often with two spines near the exterior opening.

55. **Diplograptus mucronata.** Hall. (94) 470, (104) 20.
Bulla, Victoria.

56. **Diplograptus pristis.** Hisinger. (94) 470, (104) 11.
Victoria.

57. **Diplograptus rectangularis.** McCoy. (94) 470, (104) 11.
Bulla, Victoria.

58. **Diplograptus palmeus.** Barrande. (94) 471, (105) 32.
Bendigo, Victoria.

17. Genus CLIMACOGRAPTUS. Hall.

Cells vertical free, in section sub-oval, divided from each other by deep cavities, without ornament or with a simple marginal spine ; hydrosome tapering, in section circular, or divided into two flaps ; axis prolonged beyond the distal end.

59. **Climacograptus bicornis.** Hall. (94) 471, (104) 12.
Victoria.

VI.—Sub-familia PHYLLOGRAPTINÆ. Von Lendenfeld.
PHYLLOGRAPTIDÆ. Lapworth.

Hydrosome consists of four monoserial axes, which coalesce with their dorsal sides. Sicula imbedded. The broader ends close to the proximal terminations of the hydrosomes.

18. Genus PHYLLOGRAPTUS. Hall.

Leafshaped, cells rectangular, the lateral surfaces touching. Outer touching with two protruding spines.

59. **Phyllograptus folium.** Hisinger. (94) 471, (104) 7.
Victoria.

II.—Group RETIOLOIDEA.

Lapworth.

No sicula. The cœnosarc of the common canal develops a double row of cells. Epidermis supported by chitinous fibres.

VII.—Sub-familia GLOSSOGRAPTINÆ.   Von Lendenfeld.
GLOSSOGRAPTIDÆ.   Lapworth.

Both axes united in the middle of the body.

VIII.—Sub-familia GLADIOGRAPTINÆ.   Von Lendenfeld.
GLADIOGRAPTIDÆ.   Lapworth.

Both axes separate.   Perfect exo-skeleton of chitinous fibres.

19. Genus RETIOLITES.   Barrande.

Hydrosome simple, tapering towards both ends. Axes straight or zig-zag shaped, often rudimentary, cells rectangular. Both rows alternating. Inner periderm layer a wide-meshed net.

61. **Retiolites Australis.** McCoy.   (94) 472, (105) 36.
    Keilor, Victoria.

VII. Familia PLUMULARIDÆ.   Hincks, 1868.

Hydropolypinæ forming colonies. The trophosomes possess one verticil of filiform tentacles, and are enclosed by bilateral symmetrical hydrothecæ, which are adnate to the stem and branches of the colony. Many polyps are converted into mouth and tentacleless defensive polyps. The sexual product are matured exclusively in the polypostyles.

20. Genus PLUMULARIA.   McCrady.

The Hydrothecæ cup-shaped ; the nematophores distributed along the stem and branches. Polypostyles different in the two sexes, surrounded by a simple gonangium.   (54) 529, (65) II 636, (131) 172.

I. Sub-genus.—MONOPYXIS.   Kirchenpauer.

Pinnately branched Plumularidæ ; every pinna bears a single hydranth only. Monosyphonic.

62. **Plumularia Australis.** Bale.   (9) 143, (94) 475.
    Plumularia obliqua var. Australis.   (72) 49.
    South coast of Australia (Portland).

63. **Plumularia compressa.** Bale.   (8) 43, (9) 142, (94) 475.
    South coast of Australia (Portland).

64. **Plumularia hyalina.** Bale.   (8) 41, (9) 141, (94) 475.
    South coast of Australia (Port Phillip).

65. **Plumularia pulchella.** Bale.   (8) 42, (9) 140, (94) 475.
    South coast of Australia (Port Phillip).

66. **Plumularia spinulosa.** Bale. (8) 42 (9) 139, (94) 475.
South coast of Australia (Port Phillip). East coast of New Zealand (Timaru).

67. **Plumularia obliqua.** Hincks. (9) 138, (62) 258, (63) 304, (94) 475.
Laomedea obliqua (67) 106, (123).
Campanularia (103) 372.

II. Sub-genus.—APOSTASIS. Von Lendenfeld.

The internodes of stem and branches not alternately longer and shorter. Monosyphonic. Hydrothecae not adnate.

68. **Plumularia obconica.** Kirchenpauer. (9) 127, (72) 46, (94) 473.
South coast of Australia (St. Vincent's Gulf).

69. **Plumularia producta.** Bale. (8) 39, (9) 133, (94) 474.

70. **Plumularia Buskii.** Bale. (9) 125, (94) 473.
South coast of Australia (Western Port).

71. **Plumularia tripartita.** Von Lendenfeld. (94) 477.
South coast of Australia (Port Phillip). East coast of New Zealand (Timaru).

72. **Plumularia Badia.** Kirchenpauer. (9) 128, (72) 45, (94) 473.
East coast of Australia (Brisbane).

III. Sub-genus.—HAPTOTHECA. V. Lendenfeld.

The internodes of stem and branches not alternately longer and shorter. Monosyphonic. The hydrothecae semiconic, attached in such a way that the stem forms part of the circumference of the trophosome.

73. **Plumularia cornuta.** Bale. (9) 32, (94) 474.
East coast of Australia (Port Denison).

74. **Plumularia rubra.** V. Lendenfeld. (94) 476.
East coast of Australia (Port Jackson).

75. **Plumularia Ramsayi.** Bale. (9) 131, (94) 473.
East coast of Australia (Port Denison, Albany Passage).

76. **Plumularia gracilis.** V. Lendenfeld. (94) 476.
North coast of Australia (Torres Straits).

IV. Sub-genus.—POLYSYPHONIA. V. Lendenfeld.

The stem polysyphonic.

77. **Plumularia Torresia.** V. Lendenfeld. (94) 477.
North coast of Australia.

78. **Plumularia laxa.** Allman. (6) 19, (94) 476.
36° 56' S., 150° 30', E. from Greenwich ("Challenger," 163).

79. **Plumularia campanula.** Busk. (9) 124, (18) 401, (94) 473.
South coast of Australia (Bass' Straits).

80. **Plumularia aglaophenoides.** Bale. (9) 126, (94) 473.
East coast of Australia (Broughton Island).

V. Sub-genus.—ANISOCOLA. Kirchenpauer.

The internodes are alternately longer and shorter. Hydranthes only on the longer internodes.

81. **Plumularia filicaulis.** Poeppig. (9) 134, (72) 47, (94) 474, (116).
South coast of Australia (Portland).

82. **Plumularia Goldsteini.** Bale. (8) 40, (9) 137, (94) 474.
South coast of Australia (Port Phillip, Portland).

83. **Plumularia setaceoides.** Bale. (8) 40, (9) 136, (94) 474.
East coast of Australia (Botany Bay). South coast of Australia (Port Phillip, Portland).

84. **Plumularia delicata.** Bale. (8) 40, (9) 157, (94) 474.
South coast of Australia (Western Port, Portland).

DOUBTFUL SPECIES.

85. **Plumularia scabra.** De Lamarck. (9) 145, (15) 478, (82) 164, (94) 176.
Australia.

86. **Plumularia filamentosa.** De Lamarck. (9) 144, (15) 478, (82) 164, (94) 475.
Australia.

87. **Plumularia sulcata.** De Lamarck. (9) 145, (15) 478, (82) 164, (94) 475.
Australia.

21. Genus ANTENNULARIA. De Lamarck.

Plumularidæ which possess verticillate blanchlets. Nematophores distributed along the stem and branches. (54) 529, (131) 188.

88. **Antennularia cylindrica.** Bale. (9) 146, (94) 478.
East coast of Australia (Port Curtis).

89. **Antennularia cymodocea.** Busk. (17), (9) 146, (94), 478.
Australia.

22. Genus HALICORNARIA. Bale.

Trophosomes and nematophores form groups which consist of one mesial nematophore above, and two lateral nematophores below the hydrotheca. Gonophor naked.

90. **Halicornaria ilicistoma.** Bale. (9) 184, (94) 488.
Aglaophenia ilicistoma (8) 33.
South coast of Australia (Robe, Port Phillip).

91. **Halicornaria prolifera.** Bale. (9) 183, (94) 487.
Aglaophenia prolifera (8) 34.
South coast of Australia (Port Phillip).

92. **Halicornaria humilis.** Bale. (9) 182, (94) 487.
South coast of Australia (Port Phillip).

93. **Halicornaria longirostris.** Bale. (9) 181, (94) 487.
Aglaophenia longirostris (72) 42.
Aglaophenia Thompsoni (8) 33.
South coast of Australia (from South Australia to Western Port).

94. **Halicornaria Haswelli.** Bale. (9) 180, (94) 487.
East coast of Australia (Port Curtis).

95. **Halicornaria hians.** Bale. (9) 179, (94) 487.
Plumularia hians (18) 396.
North coast of Australia (Torres Straits).

96. **Halicornaria furcata.** Bale. (9) 178, (94) 486.
East coast of Australia (Broughton Island).

97. **Halicornaria Baileyi.** Bale. (9) 177, (94) 486.
South coast of Australia (Port Phillip).

98. **Halicornaria ascidioides.** Bale. (9) 176, (94) 486.
Aglaophenia ascidioides (8) 32.
South coast of Australia (Western Port).

99. **Halicornaria superba.** Bale. (9) 176, (94) 486.
Aglaophenia superba (8) 31.
South coast of Australia (Western Port, Port Phillip).

23. Genus HALICORNOPSIS. Bale.

Hydrocaulus pinnate; hydrothecæ, with fixed anterior nematophores. Lateral nematophores absent.

100. **Halicornopsis avicularis.** Bale. (8) 26, (9) 185, (94) 488.
Aglaophenia avicularis (72) 33.
South coast of Australia (Robe, South Australia to Bass' Straits).
South coast of Tasmania. (Hobart).

101. **Halicornopsis rostrata.** V. Lendenfeld.
Azygoplon rostratum (6), 54, (94) 488.
South Coast of Australia (Bass' Straits).

24. Genus SCIURELLA. Allman.

Hydrocladia not disposed in pinnæ, but springing from many points round the circumference of chord-like stems. Gonangia situated in the axils of the hydrocladia, provided with symmetrically disposed horn-like processes, and enclosing a ramified blastyle, the branches of which are in connection with movable nematophores distributed over the surface of the gonangium.

102. **Sciurella indivisa.** Allman. (7) 26, (94) 479.

North coast of Australia (Somerset Island).

25. Genus ACANTHELLA. Allman.

Hydrocladia pinnately disposed, bearing branches terminating in simple jointed prolongations in which the places of the hydrocladia are taken by spine-like appendages.

03. **Acanthella effusa.** Allman. (7) 27, (94) 479.

Plumularia effusa (9) 129, (18) 400, (72) 46, (94) 473.

North coast of Australia (Torres Straits).

26. Genus HETEROPLON. Allman.

Hydrocladia pinnate, hydrothecal internode, with the lateral nematophores movable, and with a mesial fixed spine-like nematophore below the hydrotheca.

104. **Heteroplon pluma.** Allman. (7) 32, (94) 480.

South coast of Australia (Bass' Straits).

27. Genus DIPLOCHEILUS. Allman.

Hydrothecæ, with a duplicature of its walls, forming an external calycine envelope, which surrounds the hydrothecæ for some distance behind the orifice. Mesial nematophore in the form of a shield-like process not adnate to the hydrothecæ ; lateral nematophores absent.

105. **Diplocheilus mirabilis.** Allman. (7) 49, (94) 485.
South coast of Australia (Bass' Straits).

28. AGLAOPHENIA McCrady.

Alimentary zooids and machopolyps form groups. The former in the centre, two lateral anterior, and one mesial superior nematophore. Gonophores in corbulæ, or on modified pinnæ.

106. **Aglaophenia brevicaulis.** Kirchenpauer. (9) 171, (72) 41, (94) 484.
South coast of Australia (Port Phillip, Ballina).

107. **Aglaophenia ramulosa.** Kirchenpauer. (9) 170, (72) 41, (94) 484.
East coast of Australia (Port Lincoln).

108. **Aglaophenia Macgillivrayi.** Dale. (7) 34, (9) 170, (94) 484.
Plumularia Macgillivrayi (18) 400.

Northern limit (Louisiade Archipelago).

109. **Aglaophenia brevirostris.** Balc. (9) 169, (94) 184.

    Plumularia brevirostris (18) 397.

North coast of Australia (Cumberland Island).

110. **Aglaophenia aurita.** Bale. (9) 169, (94) 484

    Plumularia aurita (18) 397.

North coast of Australia (Cumberland Island).

111. **Aglaophenia crucialis.** Lamouroux. (9) 168, (72) 26, (83) 169, (85) 17, (94) 483.

    Plumularia brachiata (15) 478, (82) 163.

    Plumularia crucialis (15) 478.

Australia.

112. **Aglaophenia formosa.** Kirchenpauer. (9) 168, (72) 26, (94) 484.

    Plumularia formosa (17).

Australia. New Zealand.

113. **Aglaophenia delicatula.** Balc. (9) 167, (94) 484.

    Plumularia delicatula (18) 396.

North coast of Australia (Torres Straits). East coast of Australia (Port Curtis).

114. **Aglaophenia pluma.** Lamouroux. (2) IV 358, (9) 166, (40) 546 (63) 286, (72) 23, (83) 170, (94) 483.

    Pennaria pluma (112) 94.

    Plumularia cristata (82) 161, (67) 92.

    Podded Coralline (35) 13.

    Sertularia (45) 1309 (38) VII, 1, 2, (45) 1309, (103) 369, (113) 149. Australia.

115. **Aglaophenia parvula.** Bale. (8) 35, (9) 165, (94) 483.

South coast of Australia (Port Phillip).

116. **Aglaophenia ramosa.** Bale. (9) 164, (94) 482.

    Plumularia ramosa (18) 398.

North coast of Australia (Torres Straits).

117. **Aglaophenia divaricata.** Bale. (9) 162, (94) 482.

    Aglaophenia McCoyi. (8) 36.

    Aglaophenia ramosa (18) 398.

East coast of Australia (Port Jackson). South coast of Australia (from Brighton, South Australia, to Wilson's Promontory). Tasmania (George Town).

118. **Aglaophenia Huxleyi.** Bale. (9) 161, (94) 482.

    Aglaophenia angulosa (83) 166, (85) 15.

    Plumularia angulosa (15) 478, (82) 163.

    Plumularia Huxleyi (18) 395.

East coast of Australia (Port Curtis, Port Molle, Port Denison).

119. **Aglaophenia phœnicæ.** Bale.   (9) 159, (94) 482.
   Aglaophenia rostrata (72) 45.
   Plumularia phœnicœ (18) 398.
   North coast of Australia (Torres' Straits, Port Darwin). East coast of
      Australia (Holborn Island, Port Denison, Port Molle, Gloucester
      Passage).

120. **Aglaophenia longicornis.** Kirchenpauer.   (9) 157, (72) 47, (94) 481.
   Plumularia longicornis (18) 399.
   North coast of Australia (Torres Straits).   East coast of Australia (Albany
   Passage).

121. **Aglaophenia Kirchenpaueri.** V. Lendenfeld.   (94) 480.
   South coast of Australia (Western Port).

122. **Aglaophenia rubens.** Kirchenpauer.   (9) 157, (72) 48, (94) 481.
   East coast of Australia (Port Denison).

123. **Aglaophenia squarrosa.** Kirchenpauer.   (9) 156, (72) 47, (94) 481.
   East coast of Australia (Port Stephens).

124. **Aglaophenia urens.** Kirchenpauer.   (9) 155, (72) 46, (94) 481.
   East coast of Australia (Brisbane to Port Stephens).

125. **Aglaophenia plumosa.** Bale.   (8) 37, (9) 153, (94) 481.
   South coast of Australia (Port Phillip, Western Port).

DOUBTFUL SPECIES.

126. **Aglaophenia flexuosa.** Lamouroux.   (9) 172, (83) 167, (85) 16, (94)
      485.
   Plumularia flexuosa (15) 478, (82) 166.
   Indian Ocean.

127. **Aglaophenia fimbriata.** Bale.   (9) 172, (94) 485.
   Plumularia fimbriata (15) 478, (82) 163.
   Australia.

128. **Aglaophenia glutinosa.** Lamouroux.   (9) 172, (83) 171, (85) 18,
      (94) 485.
   Plumularia gelatinosa (15) 478, (82) 167.
   Australia.

29. Genus PENTANDRA. V. Lendenfeld. (88).

Plumularidæ, in which Machopolyps are as in Aglaophenia in connection
with the alimentary zooids, inasmuch as sets of them surround each hydrotheca. The polypostyles are surrounded by a corbula, as in Aglaophenia.
Each alimentary zooid is surrounded by five Machopolypes. There being
instead of the single superior Machopolype of Aglaophenia, three : a small
one, with adhesive cells only in the centre, and a pair of large Machopolypes
at the side, which consist of two parts, one with adhesive, and one with
thread cells, and which are similar to the single superior Machopolype in
Aglaophenia.

129. **Pentandra parvula.** V. Lendenfeld. (88) 355, (94) 489.

South coast of Australia (Western Port, Bass' Straits).

130. **Pentandra Balei.** V. Lendenfeld. (97) 490.

North coast of Australia (Torres Straits).

VIII.—Familia DICORYNIDÆ. Allman, 1872.

Generative zooids free swimming polyps, with two tentacles, and without a mouth, carrying two ova each. These zooids bud only on polypostyles, and never on the alimentary zooids, which have one verticil of filiform tentacles.

30. Genus DICORYNE. Allman, 1872.

Hydrocaulus, consisting of branches or simple stems, which arise at intervals from a creeping filiform hydrorhiza. Alimentary zooids fusiform, with a single verticil of filiform tentacles surrounding the base of a conical hypostom.

131. **Dicoryne annulata.** V. Lendenfeld. (94) 491.

South coast of Australia (Port Phillip).

II. SUB-ORDO.—HYDROMEDUSINÆ.
V. Lendenfeld, 1884.

Colonies of polymorphic zooids. The alimentary zooids retain the shape of polyps, whilst those in which the sexual products reach maturity are medusæ, which may become free, or remain attached to the colony, and become rudimentary and form medusostyles (4), (20), (29), (30), (50), (54), (57), (58), (59), (65), (87), (130), (131).

IX. Familia ANTHOMEDUSIDÆ. V. Lendenfeld, 1884.

= **ANTHOMEDUSIDÆ,** Haeckel, without the Cytœidæ of Haeckel.

Medusæ become free, without otoliths, with acelli at the base of tentacles. Gonads in the wall of the gastral cavity. The alimentary polyps are not invested by chitinous cups. The medusæ bud mostly on the ordinary alimentary polyps, exceptionally they are also born on peduncles, and bud direct from the hydrorhiza. Mouthless generatio polyps. Polypostyles do not occur.

I.—Sub-familia CODONINÆ. V. Lendenfeld, 1884.
CODONIDÆ. Haeckel.

Mouth simple. Gonade a simple tube. Four radial canals and unbranched tentacles, which are scattered, or in two verticils. The medusæ bud on the polyps between the tentacles.

31. Genus SARSIA. Lesson.

Codonidæ, with four equal tentacles. Membrane often very long, never cubic. Umbrella without cap on the vortex. Exumbrella smooth, without projecting nettlework. Polyps club-shaped, with scattered capitate tentacles. (54) 525, (96) (97) (125) (131) 55

132. **Sarsia radiata.** V. Lendenfeld. (95) 583, (96) (97).

East coast of Australia (Port Jackson). South coast of Australia (Port Phillip).

133. **Sarsia minima.** V. Lendenfeld. (95) 384, 98.

East coast of Australia (Port Jackson).

32. Genus DICODONIUM. Haeckel.

Codonidæ with two opposite tentacles. On the vortex of the umbrella there is a conic gallert protuberance, with axial canal. Manubrium short, scarcely projecting beyond the orifice of the umbrella.

134. **Dicodonium dissonema.** Haeckel. (50) 27, (95) 585.

Australia.

33. Genus EUPHYSA. Forbes.

Codonidæ, with three tentacles, rudiments, and one well-developed arm. Umbrella regular. Tetramer not bilateral (excepting the larger dorsal ocellar bulb). On the vortex of the umbrella there is no gallert protuberance. The polypcolonies are gymnoblastic hydroids, with two verticils of filiform tentacles, the medusæ buds at the base of the tentacles belonging to the aboral verticil. (95) 586.

East coast of Australia (Port Jackson).

II.—Sub-familia TIARINÆ. V. Lendenfeld, 1884.
TIARIDÆ. Haeckel.

Anthomedusidæ, with four broad and folded mouth-arms; with four or four pairs of gonads, four simple and broad radial canals, and unbranched tentacles. The alimentary polyps with scattered tentacles. (98).

34. Genus PANDÆA. Lesson.

Tiarinæ, with numerous tentacles in one row, abaxial acelli outside on the base of the stomach. The edges of the stomach are connected with the radial canal by four mesenteria. Four simple gonads, with smooth surface. Longitudinal lines of thread-cells in the exumbrella.

136. **Pandæa minima.** V. Lendenfeld. (98).

East coast of Australia (Port Jackson).

35. Genus TIARA. Lesson.

Tiarinæ, with numerous tentacles in one row. Abaxial acellæ outside on the base of the tentacles. No stalk to the stomach. Edges of the stomach are joined by four mesenteria, the four radial canals. Gonads, four pinnate leaves, or eight longitudinal masses, which bear irregular transverse bulges.

137. **Tiara papua.** Haeckel. (50) 58, (95) 587.

    Æquorea mitra (99) 127.

    Turris papua (2) IV 346, (39) 639, (100) 283.

Northern limit (New Guinea). Indischer Ocean.

36. Genus TURRITOPSIS. McCrady.

Tiarinæ, with numerous tentacles in one row. One ocellus inside, on the axial side of the base of the tentacles. Stalk to the stomach. No mesenteria. Gonads, four simple perradial; simple or bipartate longitudinal bulges, divided from each other by a deep furrow, with smooth surface. Mouth-flaps, with stalked nettlewarts on the margin.

138. **Turritopsis pleurostoma.** Haeckel. (50) 67, (95) 588.

Melicerto pleurostoma (115) 353.

North-west coast of Australia (De Witt's Land).

139. **Turritopsis lata.** V. Lendenfeld. (95) 588.

East coast of Australia (Port Jackson).

III.—Sub-familia MARGELINÆ. V. Lendenfeld, 1884.

Anthomedusidæ with simple or branched mouth arms. Gonad divided into four or eight marginal flaps. Alimentary polyps one verticil of filiform tentacles.

37. Genus LIZUSA. Haeckel.

Margelinæ with simple unbranched styles round the mouth, and with four perradial bundles of tentacles.

140. **Lizusa prolifera.** V. Lendenfeld. (95) 589.

East coast of Australia (Port Jackson).

38. Genus LIMNOREA. Péron.

Margelinæ with branched or otherwise complicated mouth arms (styles) and with numerous equally distributed tentacles.

141. **Limnorea triedra.** Péron. (15) 290, (31) Pl. 52, fig. 1, (95) 591, (102) III. 5, (115) 329.

Dianæa triedra (82) 505.

Limnorea proboscidea (50) 87.

South coast of Australia (Bass' Straits).

39. Genus MARGELIS. Steenstrup.

Margelinæ with ramified or composite mouth-styles, with four perradial bunches of tentacles. Stomach small, sessil. No manubrium. Mouth-styles touching at the base. The gonads do not extend to the radial canals.

142. **Margelis trinema.** V. Lendenfeld. (98).

East coast of Australia (Port Jackson).

40. Genus NEMOPSIS. L. Agassiz.

Margelinæ with branched or composed mouth-styles, and with four perradial bunches of tentacles; mouth small. The mouth styles originate at the base of the œsophagus separately. The gonads extend from the stomach edges to the radial canals.

143. **Nemopsis favonia.** Haeckel. (50) 941, (95) 591.

Favonia octonema (2) IV, 135, (15) 290, (115) 328.

Arythia octonema (81) 503.

North-west coast of Australia (Arnheim's Land).

IV.—Sub-familia CLADONEMINÆ.

CLADONEMIDÆ. Gegenbaur.

Anthomedusidæ, with branched tentacles, with 4–8 simple or branched radial canals, and four or five pair of grastral gonads. Alimentary polyps with scattered capitate tentacles.

41. Genus PTERONEMA. Haeckel.

Cladoneciminæ, with four simple radial canals, and with four perradial tentacles, which are studded with secondary tentacles or with stalked nettle-warts.

A large brooding cavity above the stomach. Four simple gonades in the wall of the stomach ; mouth with four lips. Exumbrella smooth without nettle-ribs.

144. **Pteronema Darwinii.** Haeckel. (50) 101, (95) 592.

Australia.

145. **Pteronema ambiguum.** Haeckel. (50) 102, (95) 592.

Microstoma ambiguum (99) 130, (100) 295.

Zanclea ambigua (2) IV, 344.

Northern limit (New Guinea).

X.—Familia TUBULARIDÆ. V. Lendenfeld, 1884.

The medusæ become rudimentary and remain attached to the polyps. The polyps are alimentary; no polypostyles are formed. No hydrothecæ.

I.—Sub-familia PENNARINÆ. V. Lendenfeld.

The polypes possess a distal set of capitate, and a proximal verticil of filiform tentacles   Meduso-styles broad on the gastral-wall.

42. Genus PENNARIA. Goldfuss.

Filiform tentacles with a ridge of differentiated octoderm. Colonies pinnately branched. (54) 520, (131) 121.

146. **Pennaria Australis.** Bale. (9) 45, (95) 593.

East coast of Australia (Port Jackson).

147. **Pennaria rosea.** V. Lendenfeld. (95) 594.

East coast of Australia (Port Jackson).

148. **Pennaria Adamsia.** V. Lendenfeld. (95) 595.

East coast of Australia (Port Jackson).

II.—Sub-familia TUBULARINÆ. V. Lendenfeld, 1884.

Polyps with two verticils of filiform tentacles. Meduso-styles bud on tentacular processes.

43. Genus TUBULARIA. Linné.

Simple or slightly branched, long hydrocauli. Polyps situated terminally.

149. **Tubularia Ralphi.** Halley. (9) 42, (52), (95) 596.
South coast of Australia (Port Phillip).

150. **Tubularia pygmæa.** Lamouroux. (9) 42, (15) 471, (82) 127, (83) 232, (85) 758, (95), 596.
Australia.

151. **Tubularia spongicula.** V. Lendenfeld. (95) 597.
East coast of Australia (Port Jackson).

152. **Tubularia gracilis.** V. Lendenfeld. (95) 597.
East coast of Australia (Port Jackson).

44. Genus TIBIANA. De Lamarck.

Polyps born laterally ; alternate or rarely scattered on the stems.

153. **Tibiana ramosa.** De Lamarck. (9) 43, (15) 469, (82) 206, (83) 219, (85) 743, (95) 598, (126) 425.
Australia.

IV.—Sub-familia ATRACTYLINÆ. V. Lendenfeld, 1884.

The polyps possess a single verticil of filiform tentacles. The medusoid buds are produced on the hydrocaulus.

II.—Familia LEPTOMEDUSIDÆ. V. Lendenfeld.
LEPTOMEDUSÆ. Haeckel.

Medusæ with acelli or octodermal otolithes and gonads, developed in the walls of the radial canals.
Medusæ mostly budding on transformed polypes, polypostyles,
Alimentary polyps and polypostyles invested by a chitinous perisarc. The polyps possess one verticil of filiform tentacles. (16), (24), (57) 70, (58) 22,

I.—Sub-familia THAUMANTINÆ. V. Lendenfeld, 1884.
THAUMANTIDÆ. Gegenbaur.

Leptomedusæ without marginal vesicles and simple radial canals.

45. Genus DISSONEMA. Haeckel.

Thaumantinæ with four gonads, two opposite parradial tentacles and no marginal clubs or cirri.

154. **Dissonema saphenella.** Haeckel. (50) 126, (95) 599.

c

46. Genus OCTORHOPALON. V. Lendenfeld.

Thaumantinæ with large marginal clubs in the interradii. No marginal cirri, eight tentacles, the perradial ones larger than the others. Gonangia along the whole length of the canals, join centripetally.

155. **Octorhopalon fertilis.** V. Lendenfeld. (98).

East coast of Australia (Port Phillip).

II.— Sub-familia CANNOTINÆ. V. Lendenfeld, 1884.
CANNOTIDÆ. Haeckel.

Leptomedusæ without marginal vesicles, with branched radial canal.

47. Genus CANNOTA. Haeckel.

Cannotinæ with four radial canals from each of which two branches originate, so that twelve terminations occur on each of which there is a gonad.

156. **Cannota dodecantha.** Haeckel. (50) 151, (95) 600.

Northern limit (New Guinea).

48. Genus CLADOCANNA. Haeckel.

Cannotidæ, with six dichotomously branched radial canals. All branches (about 48) extend to the ring-canal. Numerous gonads in the distal part of the branch canal.

157. **Cladocanna thalassina.** Haeckel. (50) 160, (95) 600.

  Æquorea thalassina. (81) 497.

  Berenice eachroma. (2) IV, 345, (15) 276, (31) pl. 53, fig. 2.

  Berenice thalassina. (2) IV, 345, (37) 120, (120) 327.

North coast of Australia (Arnheim's Land).

158. **Cladocanna polyclada.** Haeckel. (50) 161, (95) 601.

Northern limit (New Guinea).

III.—Sub-familia EUCOPINÆ. V. Lendenfeld, 1884.
EUCOPIDÆ. Gegenbaur.

Leptomedusæ with marginal vesicles, and four simple radial canals. Polypcolonies mostly consisting of a creeping hydrorhiza, from which either simple hydrocauli arise, or high more or less branched stems. Terminal the hydrocauli extend to form chitinous cups around the alimentary zooids.

49. Genus EUCOPE. Gegenbaur.

Eucopinæ with eight marginal vesicles, eight tentacles, no marginal cirri, no stalk to the stomach, and four gonads. Polyps with radial hydrothecæ; medusæ produced on polypostyles.

159. **Eucope annulata.** V. Lendenfeld. (95) 602.
East coast of New Zealand (Lyttelton).

160. **Eucope hyalina.** V. Lendenfeld. (98).
East coast of Australia (Port Jackson).

50. Genus OBELIA. Péron et Lesueur.

Eucopinæ, with eight marginal vesicles and numerous tentacles. Marginal vesicles at the axial side of the adradial tentacles. No marginal cirri, four gonads. Subumbrella, with rudimentary vellum. No stalk to the stomach. Polyps with radial hydrothecæ on a ramified colony. Medusæ bud on polypostyles. (54) 526, (65) II 634, (131) 155.

161. **Obelia Australis.** V. Lendenfeld. (95) 603, (98).
East coast of Australia (Port Jackson).

162. **Obelia geniculata.** Allman. (9) 59, (16) 174, (25), (63) 149, (95) 603.
Campanularia geniculata (40) 548.
Laomedea geniculata (83) 208.
Obelia lucifera (50) 175.
Sertularia geniculata (45) 1312.
Thaumantias lucifera (41) 52.

East coast of Australia (Port Jackson). South coast of Australia (from King George's Sound to Western Port). New Zealand (Lyttelton, a.o.p.).

51. Genus TIAROPSIS. L. Agassiz.

Eucopinæ with eight adradial marginal vesicles and numerous tentacles. Eight marginal vesicles with numerous otolithes, always between two tentacles. No marginal cirri. Four gonads, stomach sessil.

163. **Tiaropsis Macleayi.** V. Lendenfeld. (95) 604.
East coast of Australia (Port Jackson).

52. Genus MITROCOMIUM. Haeckei.

Eucopinæ with sixteen marginal vesicles and eight tentacles, marginal cirri between them. Four gonads. Stomach sessil.

164. **Mitrocomium annæ.** V. Lendenfeld. (95) 606.
East coast of Australia (Port Jackson).

53. Genus EUTIMALPHES. Haeckel.

Eucopinae with eight adradial marginal vesicles and numerous tentacles. Between them marginal cirri. Four gonads; a long stalk to the stomach.

165.—**Eutimolphes pretiosa.** Haeckel. (50) 195, (95) 607.
Australia.

IV.—Sub-familia EUCOPELLINÆ. V. Lendenfeld, 1883.

Medusæ without stomach and tentacles, highly developed organs of sense, eight marginal vesicles. Four radial canals, which send branches into the four gonads. Alimentary polyp in a radial hydrotheca. Polypostyle consisting of four radial tubes, between which the medusæ bud.

54. Genus EUCOPELLA. V. Lendenfeld.

Colony consisting of a hydrorhiza from which simple unbranched stems rise up. (89.)

166. **Eucopella campanula.** V. Lendenfeld. (89), (95) 607.

South coast of Australia (Port Phillip). East coast of Australia (Port Jackson).

V.—Sub-familia ACQUORINÆ. V. Lendenfeld.
= ACQUORIDÆ. Eschscholtz.

With marginal vesicles, and numerous, often branched, radial canals.

55. Genus ZYGOCANNA. Haeckel.

Stomach sessil, broad and long. Margin of mouth split into two numerous folded mouth-flaps.

167. **Zygocanna costata.** Haeckel. (50) 214, (95) 608.

Northern limit (New Guinea).

168. **Zygocanna pleuronota.** Haeckel. (50) 215, (95) 609.

Æquorea pleuronota (102) pl. XI, fig. 306, (115) 338.

Polyxenia pleuronota (37) 119.

North coast of Australia (Arnheim's Land).

56. Genus ZYGOCANNATA. Haeckel.

Æquoridæ with twelve dichotom. Radial canals crispshaped. Gonads in the shape of composed branches at the terminations of the twenty-four branches. Stomach broad and flat without stalk. Mouth wide. Margin of mouth simple, without flaps or frills.

169. **Zygocannata purpurea.** Haeckel. (50) 215, (95) 609.

Æquorea purpurea (2) IV, 360 (31), pl. 43, fig. 3 (102), pl. XI, figs. 1, 2 (115), 337.

Polyxenia purpurea Eschscholtz (37) 119.

West coast of Australia (Endrachts Land).

57. Genus ZYGOCANNULA. Haeckel.

Æquoridæ with numerous dichotome radial canals. A gonad on each branch. Stomach, at the termination of a large conic stalk, split up into mouth-flaps. The splits reach nearly to the stalk. Mouth-flaps large and folded.

170. **Zygocannula diploconus.** Haeckel. (50) 216, (93) 610.

Northern limit (Sundasea).

171. **Zygocannula undulosa.** Haeckel. (50) 217, (93) 610.

Æquorea undulosa (102) pl. XII., figs. 1-4 (115) 338.

Polyxenia undulosa (100) 314.

North coast of Australia (Arnheim's land).

58. Genus ÆQUOREA. Péron et Lesueuer.

Æquorinæ with numerous radial canals. Stomach flat, without manubrium. Lateral stomach wall rudimentary, very low. Mouth wide open ; margin of mouth simple, without flaps or frills.

172. Æquorea eurhodina. Péron et Lesueuer. (2) IV. 359, (50) 220, (95) 610, (102) pl. IX, (115) 336.

South coast of Australia (Bass' Straits).

59. Genus RHEGMATODES. A. Agassiz.

Æquorinæ with numerous simple radial canals. Stomach small, funnel-shaped, reserved, conic, constricted below. Margin of mouth simple, smooth, or crispy, without flaps or frills.

173. Rhegmatodes thalassina. Haeckel. (50) 222, 95, 611.

Æquorea cyanea (2) IV, 359, (15) 277, (102) pl. X, figs. 1-6.

Æquorea thalassina (115).

North coast of Australia (Arnheim's Land).

XII.—Familia CAMPANULINIDE.

V. Lendenfeld. 1884.

Polyp-colonies, which are differentiated into alimentary polyps, with one verticil of filiform tentacles and generative polyps, polypostyles without mouth or tentacles. Both are invested by chitinous capsules. The polypostyles only produce medusæ by budding, which remain sessil and become rudimentary meduso-styles. (10) (11) (14) (86) (131) 158.

I.—Sub-familia CAMPANULININÆ. V. Lendenfeld.

The hydrothecæ are radially symmetrical and situated terminally on the stem and branches.

60. Genus CAMPANULARIA. Hincks.

Alimentary polyps attached to the bottom of the simple cup-shaped hydrothecæ, with a broad and basal disc. The hydrocauli arise direct from the hydrorhiza. (42) (65) II, 631 (131), 144.

179. Campanularia simplex. Bale. (9) 58.

Laomedea simplex (83) 207, (85) 482.

Australia.

175. Campanularia tincta. Hincks (9) 57, (60), (93) 403.

Hincksia tincta (2).

South coast of Australia (Port Phillip, Portland).

176. Campanularia costata. Bale. (9), (93) 403.

North coast of Australia (Port Darwin).

177. **Campanularia macrocytharia.** De Lamarck. (9) 56, (82) 135, (93) 402.

Clytia macrocytharia (85) 202, (118) pl. 93, figs. 4, 5.

Australia

178. **Campanularia caliculata.** Hincks. (62) 178, (63) 164, (98).

Campanularia breviscyphia (120) 49.

Clytia poterium (2) IV, 297.

South coast of Australia (Port Phillip).

179. **Campanularia urnigera.** De Lamarck. (9) 55, (82) 135, (93) 402.

Clythia urnigera (83) 203, (85) 202.

Australia.

61. Genus HALECIUM. Oken.

The polyps are slender, they produce several hydrothecæ, one after the other at short intervals, so that they appear enclosed by several hydrothecæ one inside the other.

189. **Halecium tenellum.** Hincks. (9) 65, (61) 252, (93), 405.

Halecium labrosum. Alder (Ann. Mag. 3, III), 353.

II. Sub-familia SERTULARINÆ. V. Lendenfeld, 1884.

The hydrothecæ bilateral, symmetric, adnate to the stem and branches.

62. SERTULARIA. Hincks.

The polyps biserial, the position of the polyps in one series determined by the position of the polyps in the other. The distance between the polyps in opposite parts of the two series the same. (71) (131) 165.

I. GROUP.—APERTURE SMOOTH.

I Division.—UNBRANCHED SPECIES.

181. **Sertularia conferta.** Bale. (9) 93, (93) 412.

Dynamena conferta (71) 10.

North coast of Australia (Carpentaria Gulf).

182. **Sertularia turbinata.** De Lamarck. (9) 96, (82) 154, (93) 413.

Dynamena turbinata (83) 180, (85) 290.

II Division.—IRREGULARLY BRANCHED SPECIES.

183. **Sertularia arbuscula.** Lamouroux (9) 95, (15) 481, (82) 151, (83) 191, (85) 681, (93), 412.

Australia.

184. **Sertularia typica.** V. Lendenfeld. (93) 413.
Dynamena sertularoides (83) 173, (85) 289.
Sertularia sertularoides (71) 96.
Australia.

III Division.—DICHOTOMOUSLY BRANCHED SPECIES.

185. **Sertularia rigida.** Lamouroux. (9) 97, (15) 481, (83) 190, (85) 681, (93), 414.
Australia.

IV Division.—PINNATE SPECIES.

186. **Sertularia simplex.** V. Lendenfeld. (98).
East coast of New Zealand (Lyttelton).

187. **Sertularia tubiformis.** De Lamarck. (9) 95, (82) 153, (93) 412.
Dynamena tubiformis (15) 484, (83) 178, (95) 289.
Australia.

188. **Sertularia orthogonia.** Busk. (9) 88, (18) 390, (93) 411.
North coast of Australia (Torres Straits).

189. **Sertularia patula.** Busk. (9) 88, (18) 390, (93) 411.
South coast of Australia (Bass' Straits, Port Phillip).

II GROUP.—HYDROTHECA WITH TWO SMALL PROTUBERANCES. PINNATE.

190. **Sertularia Australis.** W. Thompson. (9) 72, (93), 408 (127) 105.
Dynamena Australis (72) 11.
South coast of Australia (Port Phillip), Tasmania (George Town), New Zealand.

191. **Sertularia penna.** Bale. (9) 74, (93) 409.
Dymonema penna (71) 11.
South coast of Australia (Bass' Straits), Van Diemen's Land.

192. **Sertularia bicornis.** Bale. (8) 22, (9) 83, (93) 410.
South coast of Australia (Port Phillip).

193. **Sertularia trigonostoma.** Busk. (9) 84, (18) 392, (93) 410.
North coast of Australia (Torres Straits), East coast of Australia (Albany Passage).

194. **Sertularia tuba.** Bale. (9) 87, (93) 411.
South coast of Australia (Port Phillip, Portland).

I Group.—HYDROTHECA WITH TWO DISTINCT TEETH.

1 Division.—UNBRANCHED SPECIES.

195. **Sertularia minima.** Thompson.   (9) 89, (93) 411, (127) 104.

    Sertularia pumila (25).

    Sertularia pumiloides (8) 21.

    Sythecium gracilis (25).

South coast of Australia (Port Phillip, Portland, St. Vincent's Gulf), New Zealand.

II Division.—IRREGULAR BRANCHED SPECIES.

196. **Sertularia irregularis.** V. Lendenfeld.   (93) 406.

South coast of Australia (Port Phillip).

III Division.—DICHOTOMOUSLY BRANCHED SPECIES.

197. **Sertularia bispinosa.** Loughtrey.   (9) 68, (25), (46), (93), 407, (127) 107.

South coast of Australia (from Brighton, South Australia, to Bass' Straits), New Zealand.

198. **Sertularia operculata.** Linné.   (9) 67, (36) 39, (38) t. IV, figs. 1, 2, (63) 263, (67) 77, (82) 144, (93) 407.

    Amphisbetia operculata (2).

    Dynamena fascicidata (71) 12.

    Dynamena operculata (15) 483, (40) 544, (83) 176, (84) 12, (85) 288.

    Sea-hair (35) 8.

    Sertularia usneoides (113) 132.

East coast of Australia (Port Stephens), South coast of Australia (from Port Elliot, South Australia, to Western Port), New Zealand.

IV Division.—PINNATE SPECIES.

199. **Sertularia loculosa.** Busk.   (9) 91, (18) 393, (93) 412.

South coast of Australia (Port Phillip, Portland, Western Port).

200. **Sertularia tenuis.** Bale.   (9) 82, (93) 410.

South coast of Australia (Port Phillip).

201. **Sertularia divergens.** De Lamarck.   (82) 153.

    Dynamena divergens (15) 484, (83) 180, (85) 290.

    Sertularia divergens (9) 81, (18) 392, (93) 410.

    Sertularia flasculus (127) 104.

North coast of Australia (Torres Straits), South coast of Australia (Port Phillip, Portland).

202. **Sertularia macrocarpa.** Bale.   (9) 80, (93) 410

South coast of Australia (Port Phillip, Portland).

203. **Sertularia unguiculata.** Busk. (9) 76, (18) 394, (26), (93) 409.
Thuiaria ambigua (127) 111.

North coast of Australia (Torres Straits), East coast of Australia (Port Jackson), South coast of Australia (from Robe, South Australia, to Western Port), New Zealand.

204. **Sertularia geminata.** Bale. (9) 78, (93) 409.
South coast of Australia (Port Phillip, Portland).

205. **Sertularia recta.** Bale. (8) 23, (9) 79, (93) 410.
South coast of Australia (Brighton, South Australia).

206. **Sertularia flexilis.** Thompson. (8) 103, (9) 78, (93) 409.
Australia.

207. **Sertularia pulchella.** Thompson. (9) 71, (93) 408, (127) 108.
South coast of Australia (South Australia), Tasmania (George Town).

208. **Sertularia bidens.** Bale. (9) 70, (93) 408.
South coast of Australia (Port Phillip).

209. **Sertularia maplestonii.** Bale. (9) 70, (93) 408.
South coast of Australia (Portland).

IV Group.—HYDROTHECA, with several teeth.

I Division—UNBRANCHED SPECIES.

210. **Sertularia Grosse-dentata.** Bale. (9) 94, (93) 412.
Dynamena Grosse-dentata (71) 13.
Australia.

211. **Sertularia minuta.** Bale. (8) 21, (9) 90, (93) 411.
South coast of Australia.

212. **Sertularia obliqua.** De Lamarck. (9) 96, (82) 154, (93) 413.
Dynamena obliqua (15) 484, (83) 179, (85) 290.
Australia.

II Division—IRREGULARLY BRANCHED SPECIES.

213. **Sertularia distans.** Lamouroux. (9) 97, (83) 191, (85) 681, (93) 414.
Australia.

III Division—DICHOTOMOUSLY BRANCHED SPECIES.

214. **Sertularia barbata.** Bale. (9) 96, (93) 413.
Dynamena barbata (15) 289, (83) 178, (85) 289.
Sertularia ciliata (82) 151.
Australia.

215. **Sertularia trispinosa.** Loughtrey. (9) 69, (25) (93) 408.
New Zealand.

IV Division.—PINNATE SPECIES.

216. **Sertularia scandens.** Lamouroux. (9) 97, (15) 481, (83) 189, (85) 681, (93) 414.

   Sertularia millefolium (82) 141.

   Australia.

217. **Sertularia tridens.** V. Lendenfeld.

   Sertularia tridentata (9) 97, (82) 151, (83) 187, (85) 680, (93) 414.

   Australia. (East coast of Australia, Port Jackson).

218. **Sertularia crenata.** Bale. (9) 86, (93) (411).

219. **Sertularia insignis.** Thompson. (9) 86, (93) 410, (127) 109.

   Tasmania (George Town).

220. **Sertularia acanthostoma.** Bale. (8) 23, (9) 85, (93) 410.

   South coast of Australia (South Australia).

221. **Sertularia tridentata.** Busk. (9) 79, (18) 394, (93) 409.

   South coast of Australia (Bass' Straits).

222. **Sertularia elongata.** Lamouroux. (9) 75, (15) 481, (83) 189, (85) 681, (93) 409, (127) 107.

   Dynamena abietinoides (46).

   Sertularia abietinoides (25).

   Sertularia lycopodium (82) 142.

   South coast of Australia (Bass' Straits, Port Phillip, South Australia), New Zealand.

223. **Sertularia divaricata.** De Lamarck. (82) 143.

   South coast of Australia (Bass' Straits).

224. **Sertularia fertilis.** V. Lendenfeld. (93) 406.

XII.—Familia HYDRACTINIDÆ.

V. LENDENFELD.

= HYDRACTINIDÆ. Claus? (23).

Polyp-colonies producing free medusæ or medusostyles. The free medusæ possess acelli at the base of the tentacles, and no atolithes. The tentacles are scattered, and equally distributed. With simple mouth arms. The polyp-colonies consist of a dense mass of entwined hydrorhiza, from which the hydrocauli, simple, or slightly branched, grow forth. The alimentary polyps possess one verticil of filiform tentacles. The medusæ bud generally from the hydrorhiza. Polypostyles. Besides the alimentary and generative polyps we meet with defensive spiral polyps. (117) (131) 67.

I.—Sub-familia CYTÆINÆ. V. Lendenfeld, 1884.

Producing free medusæ (48), (54) 517, (65), II., 621.

II.—Sub-familia HYDRACTININÆ. V. Lendenfeld, 1884.

Sexual generation. Medusostyles.

63. Genus DEHITELLA. Gray (?).

Hydrophyton dichotomous, perisarc dark-brown. Perisarc of spiral zooids vertical on the stem and branches (?) Differs from Ceratella by the equal distribution of the tufts of spicules. Spiralzooids ?

225. **Dehitella atrorubens.** Gray. (9) 49, (19) 47, (95) 612.

East coast of Australia (Bondi Bay).

64. Genus CERATELLA. Gray (?).

Hydrophyton irregularly dichotomous. Perisarc dark-brown. Hydranths, polypostyles, and medusoid buds unknown. The detail description of the skeleton by Gray is worthless.

226. **Ceratella fusca.** Gray. (9) 48, (19) (47) (95) 612.

Australia.

III. SUB-ORDO.—HYDROCORALLINÆ.

Moseley, 1876.

Alimentary zooids, with few verticillate, capitate tentacles. Hydrorhiza forming a dense calcareous skeleton, which also invests the polyps. Groups of machopolyps, in the form of tentacular zooids, surround the alimentary polyps. Ultimately generative zooids are probably medusæ. (2) (111).

XIV. Familia STROMATOPORIDÆ.

Murie and Nicholson, 1878.

Possessing undulating laminæ in the skeleton.

XV. Familia MILLEPORIDÆ.

Moseley, 1877.

Alimentary zooids, with four to six tentacles. Polypary with many conic spaces, divided by tabulæ. Cœnenchym, with reticulating canals. Dactylo-zooids, with numerous tentacles. Ampullæ absent.

65. Genus MILLEPORA. Moseley.

In the centre of each irregular group of dactylozooids one alimentary polyp. 110.

227. **Millepora tortuosa.** Dana. (27) 515, (95) 613.

Fiji Islands.

66. Genus ARACHNOPORA.   Tenison-Woods.

The colony spreading like a small thin web over corals.

228. **Arachnopora argentea.**   Tenison-Woods.   (95) 614, (131) 8.
Australia.

XVI. Familia STYLASTERIDÆ.
Moseley.

Alimentary zooids, with from four to twelve tentacles. Massive hydrosome containing tubes which possess pseudosepta, formed by the regular position of the tentacular zooids.   Dactylozooids devoid of tentacles.   Gonanzia contained in ambulæ.

67. Genus DISTICHOPORA.   De Lamarck.

Pores sporadic; dactylopores of one kind only.   Pores simple, in a triple linear row at the lateral edges of the branches of the flabellum, rarely on its faces.

229. **Distichopora violacea.**   Pallas.   (95) 614, (113) (133) 61.
North coast of Australia.

230. **Distichopora rosea.**   Saville Kent.   (69) 281, (95) 614, (133).
East coast of Australia.

231. **Distichopora coccinea.**   Gray.   (95) 615, (130).
North coast of Australia.

232. **Distichopora gracilis.**   Dana.   (27) 151, (95) 615, (133).
North coast of Australia.

233. **Distichopora livida.**   Tenison-Woods.   (95) 615, (133).
North coast of Australia.

68. Genus STYLASTER.   Gray.

Pores occurring in regular cyclo-systems.   Both kinds of pores with styles. Corallum increasing by regular alternate gemmation of the cyclo-systems from one another.   Gasterozooids with eight tentacles.

234. **Stylaster gracilis.**   Milne Edwards et Haime.   (32) 98, (33) (95) 615.
Australia.

235. **Stylaster sanguinens.**   Valenciennes.   (32) 96, (33) (95) 615.
Australia.   New Zealand.

236. **Stylaster gemmascens.**   Milne Edwards et Haime.   (33) 130, (95) 615.
Indian Ocean.

69. Genus **CRYPTOHELIA**. Milne Edwards.

Pores in regular cyclo-systems. Styles absent. Gastrozooids without tentacles; gastrophores with two chambers; summits of the cyclo-system covered by a lid.

237. **Crytophelia pudica.** Milne Edwards et Haime. (32) 93, (33) (95) 616.

Northern limit (New Guinea).

IV. SUB-ORDO.—TRACHOMEDUSINÆ.

V. Lendenfeld, 1884.

TRACHOMEDUSÆ. Claus.

Hydromedusæ which are medusæ, and are propagated sexually without a change of generation, and without forming polypoid zooids. Medusæ, with entodermal acustic clubs.

XVII.—Familia TRACHOMEDUSIDÆ.

V. Lendenfeld, 1881.

= **TRACHOMEDUSÆ.** Haeckel.

Trachomedusinæ with gonads on the radial side. (34) (50) (51) (57) (58) (59).

I.—Sub-familia PETASINÆ. V. Lendenfeld, 1884.

With four radial canals and long tube-shaped stomach.

II.—Sub-familia TRACHINEMINÆ. V. Lendenfeld, 1884.

= **TRACHINEMIDÆ.** Gegenbaur.

With eight radial canals, with long tube-shaped stomach.

III.—Sub-familia AGLAURINÆ. V. Lendenfeld, 1884.

= **AGLAURIDÆ.** L. Agassiz.

With eight radial canals and with a pedicle to the stomach.

70. Genus STAUROGLAURA. Haeckel.

Aglaurinæ with four gonads in connection with alternating radial canals, the other four canals sterile. Gonads on the stalk of the stomach, not on the sub-umbrella. The four acustic clubs in the interradii.

238. **Stauraglaura tetragonima.** Haeckel. (50) 277, (95) 617.

Australia.

IV.—Sub-familia GERGONINÆ.   V. Lendenfeld, 1884.

= GERGONIDÆ.   Haeckel.

Four or six radial tubes; leaf-shaped gonads.

Long stomach pedicles, eight or twelve marginal peronial, and as much acustic vesicles. (34), (57), 48.

71. GERGONIA.   Péron et Lesueur.

Gergonidæ with six gonads in the vicinity of the six radial canals, without centripetal canals.   Only six permanent, hollow, perradial tentacles, and twelve acustic clubs.

239. **Gergonia dianæa.**   Haeckel.   (50) 295, (95) 618.

Dianæa endrachtensis (118) 566.

Dianæa gaberti (15) 289.

Eirene endrachtensis (37) 94.

Arythia viridis (2), IV., 363.

West coast of Australia (Endracht's Land).
Indian Ocean.

72. GENUS CARMARIS.   Haeckel.

Gergonidæ, with six gonads in connection with the six radial canals, between which cœcal canals run centripetally from the ring-canal.   Twelve permanent tentacles.   (Six hollow perradial ones, and six solid interradial ones.)   Twelve auditory vesicles (six primary interradial and six secondary perradial).

240. **Carmaris.**   Giltschii.   (50) 296, (98) 618.

Australia.

XVIII.—Familia NARKOMEDUSIDÆ.

Von Lendenfeld, 1884.

= NARKOMEDUSÆ.   Haeckel.

Trachomedusæ, with gastral gonads.

I.—Sub-familia CUNANTHINÆ.   V. Lendenfeld.

= CUNANTHIDÆ.   Haeckel.

With broad, pouch-shaped radial canals, with otoporpa.

II.—Sub-familia PEGANTHINÆ.   V. Lendenfeld.

PEGANTHIDÆ.   Haeckel.

Without radial canals, and without gastral pouches in the sub-umbrella. With otoporpa.

III.—Sub-familia ÆGININÆ. V. Lendenfeld.

= ÆGINIDÆ. Gegenbaur.

Circular canal in communication with gastral cavity by double peronial tubes. With gastral pouches without otoporpa.

73. Genus ÆGINURA. Haeckel.

Æginidæ, with eight peronial double canals and eight tentacles (four perradial and four interradial), and with sixteen internemal genital pouches.

241. **Æginura myosura.** Haeckel. (50) 343, (51) 35, (95) 619.

47° 25′ S. ; 130° 32′ E. Greenwich.

IV.—Sub-familia SOLMARINÆ. V. Lendenfeld.

SOLMARIDÆ. Haeckel.

Without circular canal, and without otoporpa ; with or without radial canal.

Sydney : Charles Potter, Government Printer.— 1887.

PUBLICATIONS OF THE AUSTRALIAN MUSEUM.

Catalogue of the Specimens of Natural History and Miscellaneous Curiosities in the Australian Museum, by J. Roach. 1837. 8vo, pp. 71. (Out of print.)

History and Description of the Skeleton of a new Sperm Whale in the Australian Museum, by W. S. Wall. 1851. 8vo, pp. 66, with 2 plates. (Out of print, but to be reprinted.)

Catalogue of Mammalia in the Collection of the Australian Museum, by G. Krefft. 1864. 12mo, pp. 133. (Out of print.)

Guide to the Australian Fossil Remains in the Australian Museum. 1870. 8vo. (Out of print.)

Catalogue of the Minerals and Rocks in the Collection of the Australian Museum, by G. Krefft. 1873. 8vo, pp. xvii-115. (Out of print.)

Catalogue of the Australian Birds in the Australian Museum, by E. P. Ramsay. Part I, Accipitres. 1876. 8vo, pp. viii-64. Boards, 2s.; cloth, 3s.

Catalogue of the Australian Stalk and Sessile-eyed Crustacea, by W. A. Haswell. 1882. 8vo, pp. xxiv-324, with 4 plates. Wrapper, 10s. 6d.

Guide to the Contents of the Australian Museum. 1883. 8vo, pp. iv-56. Wrapper, 3d.

Catalogue of the Library of the Australian Museum. 1883. 8vo, pp. 178. Wrapper, 1s. 6d., with two supplements.

Catalogue of a Collection of Fossils in the Australian Museum, with Introductory Notes, by F. Ratte. 1883. 8vo, pp. xxviii-160. Wrapper, 2s. 6d.

Catalogue of the Australian Hydroid Zoophytes, by W. M. Bale. 1884. 8vo. pp. 198, with 19 plates. Wrapper, 3s. 6d.

List of Old Documents and Relics in the Australian Museum. 1884. 8vo, pp. 4.

Descriptive Catalogue of the General Collection of Minerals in the Australian Museum, by F. Ratte. 1885. 8vo, pp. 221, with a plate. Boards, 2s. 6d. ; cloth, 3s. 6d.

Catalogue of the Echinodermata in the Australian Museum, by E. P. Ramsay. Part I, Echini. 1885. 8vo, pp. iii ii-54. 5 plates. Wrapper, 2s. 6d. ; cloth, 3s. 6d.

Descriptive List of Aboriginal Weapons, Implements, &c., from the Darling and Lachlan Rivers. 1887. 8vo, pp. 8.

Notes for Collectors. 1887. 8vo, 1s. Containing
Hints for the Preservation of Specimens of Natural History. By E. P. Ramsay. Pp. 17.

Hints for Collectors of Geological and Mineralogical Specimens, by F. Ratte. Pp. 26, with a plate. (The two parts may be obtained separately. Price, 6d. each.)

Descriptive Catalogue of the Medusæ of the Australian Seas. Part I, Scyphomedusæ. Part II, Hydromedusæ. By R. von. Lendenfeld. 1887. Pp. 32 and 49. Boards, 2s. 6d. ; cloth, 3s. 6d.

In course of preparation :-

Catalogue of the Library. Revised and corrected.

Guide to the Museum. New edition.

In the press :—

Catalogue of Sponges, by R. von Lendenfeld.

Check List of Australian Birds, showing the distribution of all the known species with notes, by E. P. Ramsay.

Catalogue of Shells. Hargraves, and General Collections, by J. Brazier.

Descriptive Catalogue of the Eggs of Australian Birds, by A. J. North.

LIST OF ERRATA

IN THE

CATALOGUE OF THE AUSTRALIAN

SCYPHOMEDUSÆ AND HYDROMEDUSÆ.

BY

R. von LENDENFELD, Ph.D., F.L.S., &c.

This Catalogue was printed after Dr. von Lendenfeld had left New South Wales, and he was therefore unable to supervise the proofs. In consequence of this the mistakes made by the printers were not rectified. Dr. von Lendenfeld has now compiled the following list of principal errata.

PART I.--SCYPHOMEDUSÆ.

	For	Read
Throughout :—	CTENAPHORÆ	CTENOPHORÆ.
	Cnidaria	Cnidaria,
	laps	flaps
	Phasellatæ	Phacellotæ.
	Strabila	Strobila.
	V. Lendenfeld	v. Lendenfeld.

	For	*Read*

Page 5, first line of small type, under Classis IV. & VI. :—

                       Tree   Free.

Page 6, line 6 :—   serata   serota.

Page 7, line 2 :—   amusing   nursing.

      line 15 from bottom :—

                  lubepithel   subepithelium.

Page 8, line 5 :—   swins   swims.

      line 13 :—   radius   radii.

      line 18 :—   entremity   extremity.

      line 5 from bottom :—

                  Cubamedusæ   Cubomedusæ.

Page 9, lines 4 & 8 :—Rhizostamidæ   Rhizostomidæ.

Page 10, lines 4, 5, & 7 :—

                  autimer   antimer.

      line 9 :—   Bolyclonidæ   Polyclonidæ.

      line 14 from bottom :—

                  Euhomedusæ   Cubomedusæ.

      lines 7 & 11 from bottom :—

                  ganglia   ganglion.

Page 11, line 20 :— oval part. In all the vital   oral part. All the vital
                  organs are situated the   organs are situated in the
                  sub-umbrella.   sub-umbrella.

      line 5 from bottom :—

                  lamellons   lamellous.

                  membranes   membranous.

    omit footnote.

Page 12, line 2 :—   Schirmhählen   Schirmhöhlen.

      line 3 :—   parochial   brachial.

      line 5 :—   Chamoostomidæ   Chaunostomidæ.

      line 6, omit V.

	For	*Read*
Page 13, line 2 :—	*Lancaster*	*Lankester.*
line 5 :—	STAMOMEDUS.E	STAUROMEDUS-E.
line 15 :—	mesagan	mesogon.
Page 14, line 4, add :—on *Macrocystis.*		
line 5 :—	Macrocystis	unknown.
Page 15, line 10 :—	Peramedusæ	Peromedusæ.
line 16 :—	caves	cavities.
Page 16, lines 1 & 12 :—		
	CUBAMEDUS.E	CUBOMEDUS.E.
lines 10 & 17 :—		
	CHORYBDEIDÆ	**CHARYBDEIDÆ.**
Page 17, line 1 :—	**DISCAMEDUSÆ**	**DISCOMEDUSÆ.**
line 13 :—	**CARMASTOMÆ**	**CANNOSTOMÆ.**

line 8 from bottom, add :—Proceedings of the Linnean Society of New South Wales, vol. ix. page 206.

Page 18, line 8 :—	primary tentacles	primary radii.
line 10 from bottom :—		
	Couthony	Couthouy.
line 2 from bottom :—		
	T. de Lamarck	J. de Lamarck.
Page 19, line 15 from bottom :—		
	Discamedusæ	Discomedusæ.
line 11 from bottom :—		
	STEMOPTYDIA.	STENOPTYCIIA.
Page 20, line 15, omit :—of.		
line 13 from bottom :—		
	Local colour. Varieties of	Local colour-varieties of.
Page 21, line 17 :—	Lemostomæ	Semostomæ.
line 20 :—	(schirmhahlen)	(Schirmhöhlen).

	For	*Read*

Page 21, line 14 from bottom :—

   system   System.

  line 9 from bottom :—

    T. de Lamarck   J. de Lamarck.

Page 22, lines 5, 6, 18, & 30 :—

   cærulea   cœrulea.

Page 23, line 6 :— **CHAMIOSTOMIDÆ**   **CHAUNOSTOMIDÆ.**

  line 15 :— Centrifugally   Centripetally.

  line 21 :— band   Band.

Page 24, line 4, omit :—Glenelg, Haacke.

  lines 6–8 :—*Pseudorhiza Haeckeli*, W.   *Pseudorhiza Haeckeli*, W.
Haacke. Pseudorhiza   Haacke. Pseudorhiza
Haeckeli spec. nov. der   Haeckeli spec. nov. Der
endspross des discomedu-   Endspross des Discomedu-
senstammes biologisches   senstammes. Biologisches
centralblatt.   Bond 4,   Centralblatt.   Band 4,
Nr. 10. Seite 291.   Nr. 10, Seite 291.

  line 13, omit :—in the above loc.

  line 15, omit :—one.

Page 26, line 16 :— Parochial   Brachial.

Page 27, line 2 :— primate   pinnate.

  line 6 from bottom :—

    derided   derived.

  line 4 from bottom :—

    system der medusen   System der Medusen.

Page 28, line 16 :— Euligneital   subgenital.

Page 29. line 7 :— *Phillorhiza*   *Phyllorhiza.*

  line 3 from bottom, omit V.

	For	*Read*

Page 30, line 16 :— finally   finely.

    line 2 from bottom :—

        Neselljellen   Nesselzellen.

Page 31, line 16 from bottom :—

        SYMBIATICA   SYMBIOTICA.

    line 15 from bottom :—

        zooxanthella   zooxanthellæ.

Page 32, line 3 from bottom :—

        Australian 26 species   26 Australian species.

PART II.—HYDROMEDUSÆ.

Page 4, line 9 from bottom :—

        Tahrbuch   Jahrbuch.

    line 5 from bottom :—

        Tenaische Zeitschrift für   Jenaische Zeitschrift für

        naturwissenschaft   Naturwissenschaft.

Page 6, line 10 :— Fübingen   Tübingen.

Page 7, line 12 : — Tenaische   Jenaische.

    line 16 from bottom :—

        coelenteraten   Coelenteraten.

Page 8, line 2 :— Actimien   Actinien.

    line 21 :— seites   Seite.

    line 8 from bottom :—

        Lamaroux's   Lamouroux's.

Page 9, line 16 :— einige   einige.

    line 22 :— coelenteraten   Coelenteraten.

For	Read
Page 10, line 8 :— nessolzellon	Nesselzellen.
Page 11, line 10 :— natur	Natur.
line 9 from bottom :—	
introcelluläre	intracelluläre.
Page 15, line 8 from bottom :—	
obigactis	oligactis.
Page 16, line 8 :— clavinæ	Clavinæ.
line 10 from bottom, first word :—	
polypostyle	polypostyles.
Page 17, line 24 :— antipathies	antipathes.
line 2 from bottom :—	
Sarsii	Lairii.

Page 18, add to the synonyms of **Lafoëa fruticosa** :—*Campanularia fruticosa* (45).

Page 21, line 13 :— *tetrocytharia*	*tetracytharia.*
Page 22, line 10 :— 14. Genus Didymograpsus.	14. Genus DIDYMOGRAP-SUS.
line 19 :— octabrachiatus	octobrachiatus.
Page 23, line 21 :— palmeus	palmens.
Page 25 :— III. Subgenus.—Hapto-THECA.	III. Subgenus.—HAPTO-THECA.

Page 29, line 17 from bottom, insert after *Sertularia* the word *pluma*.

Page 32, insert :—135. **Euphysa australis** (95) 586. v. Lendenfeld—after the diagnosis of the Genus EUPHYSA.

Page 34, line 3 :— *Arythia*	*Orythia.*
Page 35, line 11 :— spongicula	spongicola.
Page 36, line 15 from bottom :—	
eachroma	*euchroma.*
Page 37, line 7 from bottom :—	
Eutimolphes	Eutimalphes.

For	*Read*
Page 38, line 4 :— campanula	campanularia.
lines 19 & 24 :—	
ZYGOCANNATA.	ZYGOCANNOTA.
Page 42, line 19 from bottom :—	
fisciculata	*fasciculata.*
line 5 from bottom :—	
flasculus	*flosculus.*
Page 43, line 19 from bottom :—	
Grosse-dentata	grosse-dentata.
Page 48, repeatedly :—	
GERGONIA, GER-GONIDÆ, &c.	GERYONIA, GERY-ONIDÆ, &c.